MARS

MARS SOCIETY DESERT RESEARCH STATION
MAJOR SPONSORS
E MUSK FOUNDATION

MARS

SCIENCE FICTION TO COLONIZATION

For general information on our other products and services or to obtain technical support, please contact our Customer Care Department within the United States at (866) 744-2665, or outside the United States at (510) 253-0500.

Lightning Guides publishes its books in a variety of electronic and print formats. Some content that appears in print may not be available in electronic books, and vice versa.

ISBN Print 978-1-942411-41-3
eBook 978-1-942411-42-0

Front Cover Photo: © DreamPictures/Getty Images. Back Cover Photo: © Antonio M. Rosario/Tetra Images/Corbis. Interior Photos: © David Howells/ Corbis, p.1; George Frey/Stringer/Getty, p.2; New York Public Library/Getty Images, p.5; Dja65/ Shutterstock, p.6;Sergei Remezov/Reuters/Corbis, p.7; LironPeer/iStock, p.9; Sfio Cracho/Shutterstock, p.9; Antonio M. Rosario/Tetra Images/Corbis, p.9; sergign/Shutterstock, p.9; Mark Garlick Words & Pictures Ltd/Science Photo Library/Corbis, p.12; © skodonnell/iStock, p.13;Denis Tabler/Shuttersock, p.14; Hulton Archive/Stringer/Getty Images, p.16; Mary Evans Picture Library/Alamy, p.19; Silver Screen Collection/Contributor, p.19; Buyenlarge/Getty Images, p.19, p.21; Fine Art Images/Heritage Images/ Getty Images, p.21; Transcendental Graphics/Getty Images, p.21; "TheLeftHandOfDarkness1stEd"[Fair Use]/Wikipedia, p.22; Universal History Archive/UIG/ Getty Images, p.25; CSA ASC LOG[Public domain]/ Wikimedia Commons, p.27; Fine Art Images/ Heritage Images/Getty Images, p.28; RIA Novosti archive, image #510848 / Alexander Mokletsov / CC-BY-SA 3.0 [CC BY-SA 3.0]/Wikimedia Commons, p.30; NSSDC, NASA[1] [Public domain]/Wikimedia Commons, p.30; NASA Glenn Research Center (NASA-GRC) (C-1964-71394) [Public domain]/Wikimedia Commons, p.31; Omikron/Getty, p.31; Sovfoto/ UIG/Getty Images, p.31; Andrzej Mirecki [CC BY-SA 3.0]/Wikimedia Commons, p.32; [Public domain]/ Wikimedia Commons, p.32; NASA/JPL [Public domain]/Wikimedia Commons, p.32; Public domain]/ Wikimedia Commons, p.32; Uwe W [Public domain], via Wikimedia Commons, p.33; Ruffnax (Crew of STS-125) derivative work: Quibik (HST-SM4.jpeg) [CC0]/ Wikimedia Commons, p.33; NASA/MSFC [Public domain]/Wikimedia Commons, p.33 Mike Licht/Flickr, p.33; AF archive/Alamy, p.35; CBS Photo Archive/ Getty Images, p.36; George Frey/Getty Images, p.38; Stocktrek Images/Getty Images, p.44; SERGEI REMEZOV/Reuters/Corbis, p. 46; [Public domain]/ Wikimedia Commons, p.49; NASA [Public domain]/ Wikimedia Commons, p. 50; Jack Weir [Public domain]/Wikimedia Commons, p.51; © CORBIS, p.51; CBW/Alamy, p.52,53; Universal History Archive/ UIG via Getty Image, p.54; ASA/JPL-Caltech/MSSS/ TAMU/Science Source, p.62,63; Astrobobo/iStock, p.65; Universal Images Group Limited/Alamy, p.66; Steve Austin/Papilio/Corbis, p.67; Radius Images/ Corbis, p.68; MyLoupe/UIG/Getty Images, p.71; Rob Atkins/iStock, p.72; George Frey/Getty Images, p.74; James A. Sugar/National Geographic/Getty Images, p.77; SERGEI REMEZOV/AFP/Getty Images, p.78; Oleg Moiseyenko, p.80; NASA/Science Source, p. 82; Marc Ward/Shutterstock, P.84; Mark Williamson/ Getty Images, p.87; Africa Studio/Shutterstock, p.89; Greatstock/Barcroft Media, p.91; GoodGnom/iStock, p.92; Bryan Bedder/Getty Images for Engadget Expand, p.95; SERGEI REMEZOV/Reuters/Corbis, p.96; Science Photo Library/Corbis, p.101; NASA/ Newsmakers/Getty Images, p.103; © ZUMA Press, Inc/Alamy, p.106

FROM THE EDITOR

"I would like to die on Mars. Just not on impact."

—ELON MUSK

Blood red in the night sky, symbol of the god of war, Mars has captured the imagination of humans since ancient times. Early astronomers envisioned canals on the surface, strange waterways where extraterrestrial Argonauts poled wondrous barges. Ray Bradbury's *The Martian Chronicles* thrilled readers with alien encounters in a beautiful and hostile world. Even Bugs Bunny grappled with the insidious Marvin the Martian and his plan to annihilate the Earth, which blocked his view of Venus. Today, rovers crawl the dusty valleys of the Red Planet, while governments and private corporations contemplate the manned exploration, colonization, and exploitation of our most similar planetary neighbor. And life on Mars, a Victorian fancy deposed by modern science, is now once again at the forefront of popular and scientific thought, albeit on a more microscopic scale. What is it about Mars that so enthralls us? That we find so riveting and irrepressible? The answer is as elusive as the mysteries of Mars itself.

CONTENTS

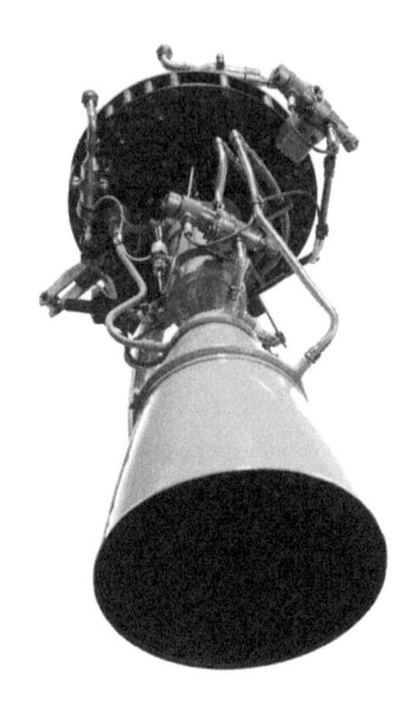

INTRODUCTION

Is Mars just a dream? After several probes and lander missions, humankind is closer than ever to colonizing our planetary neighbor. NASA plans to send a manned spacecraft to orbit Mars in the 2030s; other space agencies and private foundations are gearing up for their own missions. Still, skeptics say that anyone who believes humans will reach Mars in our lifetime must be dreaming.

In a way, the skeptics are right. Dreaming is what brought us this close to Mars in the first place. Ancient Egyptians, Greeks, and Romans told their own stories about the Red Planet. In the Victorian era, great minds believed that if any nearby planet harbored alien beings, it had to Mars. Many of the technologies that are putting Mars within our reach were first dreamed up in sci-fi novels, Saturday morning cartoons, and movie theaters. Everything we have learned about Mars began as speculation, conjecture, and bold feats of the imagination.

Today, with better space photography and more sophisticated technologies for detecting signs of potential life, we are re-imagining Mars once again. Mars has been suggested as a possible site for living extraterrestrial organisms—or at least as a place that once, very likely, harbored some form of life. Before we can answer the biggest question in the universe—Are we alone?—we need to understand our neighbor Mars: its science, its history, and most importantly, its place in our dreams.

FIRST, A FEW FACTS

Egyptian astronomers noted
MARS'S APPARENTLY
RETROGRADE MOTION
IN THE 16TH CENTURY BCE

A YEAR
ON MARS IS
687 DAYS
IN EARTH TIME

MARS'S RED COLOR
is the result of large amounts of
IRON OXIDE
ON THE SURFACE

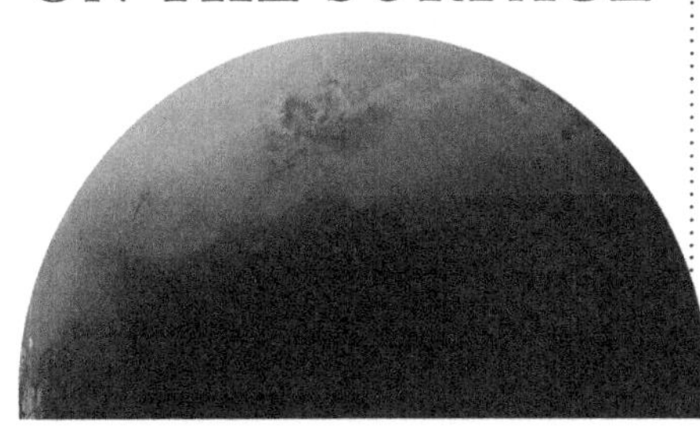

A DAY ON
MARS IS
24
HOURS
37
MINUTES

How long would I survive on Mars without a space suit?

About 60 seconds. You'd be very cold, but the cold wouldn't kill you—the low atmospheric pressure would. First your sweat and saliva would evaporate. Then you would swell up as the water in your body turned to gas. In about 30 seconds you would pass out. Your skin and organs would rupture. Death on Mars would be quick and painful. If you survived past a minute, you would suffocate, and if you survived even longer, your cells would mutate due to the radiation. This is assuming you can avoid Mars's famous dust storms, which last for a month and cover the entire planet.

Why is Mars a good candidate for extraterrestrial life?

There is evidence Mars once had liquid water, evaporated now, due to the thin atmosphere. Even today, the poles still harbor water in the form of ice. The presence of water gives strong indication that the planet might have supported simple organisms, even if it does not currently foster life.

Why haven't we sent humans to Mars already?

Space travel is dangerous and costly, even for short distances. Rovers and observation satellites are considerably more cost-effective because they don't need food and fuel (other than sunlight) once they arrive, and plans don't have to be made to bring them home again. It might have been possible—though incredibly dangerous—to send a manned mission to Mars before now, but it wasn't practical.

Why is there no liquid water on Mars?

The low atmospheric pressure on the surface of Mars does not allow for liquid water. Many geological indications, however, point to the likelihood that Mars once had liquid water and, therefore, an atmosphere dense enough to support it.

What evidence is there for liquid water in Mars's past?

A variety of surface features in Mars's geology are indicative of running water. There might be alternate explanations, but features that appear to be outflow channels, gullies, and alluvial fans support the theory of water erosion. The presence of hematite and goethite on Mars—minerals that form in the presence of water on Earth—also indicates liquid water was once a part of the planet's past.

Why is Mars the next logical target for a manned mission?

Although Venus is closer, its surface conditions—which include temperatures of at least 752 degrees Fahrenheit and an atmosphere 90 times denser than Earth's—render it extremely inhospitable. Mars is nearly as close, but it is more like our home planet, with a surface pressure equivalent to 0.6 percent of that on Earth, making it much more welcoming to human explorers.

Is it true that a pyramid and a carved face were found on Mars?

In 1976, the Viking 1 orbiter photographed the Cydonia area of Mars, revealing a geological feature that looked like a human face. Other images included pyramidal features. The cameras on the Viking orbiters were extremely low-resolution by today's standards. More recent, higher-resolution pictures from other missions show that the "face" and "pyramids" are, in fact, natural formations. A psychological phenomenon called "pareidolia" causes the human brain to perceive faces and other familiar shapes or objects where they don't actually exist.

WHY MARS?

BUILDING AN OFF-EARTH COLONY ON THE RED PLANET

For centuries, we've referred to our home planet as "Mother Earth." It's the only home we've ever had, fostering countless generations of humanity. Why, now, are we so ready to leave? Perhaps we're like young adults, grown up and ready to leave the shelter of our mother's home. Perhaps the doomsayers among us are right—the end is nigh, either by our own making or by a cataclysm we're unable to foresee. The fact is that we have limited natural resources, so perhaps it's time for some of us to see what else the universe can provide.

Whatever our reason for leaving Earth—whether temporarily on a scientific mission or permanently to establish a colony—the next logical question is "Where?" Our own moon might seem like a good candidate; after all, it's close and geologically stable.

There have been various plans—both logical and outrageous—for a moon colony, beginning as early as the 17th century. Ideas centering on moon colonization reached their pinnacle in the 1950s and '60s, not coincidentally when science fiction was transitioning out of its sensationalist pulp phase and into a more mature, thoughtful form of literature.

Although the moon would make a good location for a research base, it is less appealing as a place to colonize—a place to which humanity might retreat to create a lasting society if things go badly on Earth, or if our population continues to rise. The moon's low gravity, relative lack of atmosphere, relative lack of a variety of other necessary elements (e.g., carbon, nitrogen, and hydrogen), and extremely long lunar night count among its drawbacks.

What about Venus? Sometimes referred to as Earth's sister planet because of its proximity as well as its similar size and mass, Venus also has a terrestrial composition much like that on Earth. Our sister planet, though, has never been a serious candidate for colonization, at least not since scientists discovered its greenhouse-gases-induced high

SMALL STEPS

Although three astronauts made the Apollo 11 flight to the moon, only two landed on the surface. Commander Neil Armstrong and lunar module pilot Buzz Aldrin entered lunar module Eagle *and made the descent to the moon's surface. Command module pilot Michael Collins remained in orbit to await his crewmates' return.*

DID YOU KNOW

The earliest known use of the term *Mother Earth*, in the form of "Mother Gaia," was in a Mycenaean Greek inscription from the 13th or 12th century BCE, though myths of goddesses personifying nature existed much earlier.

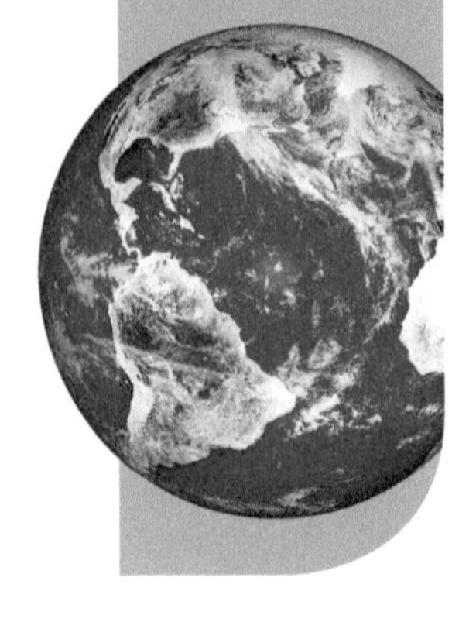

temperatures (which reach upward of 400 degrees Celsius at the surface).

Other planets in our solar system have been similarly ruled out; they're gaseous or too far away or completely inhospitable to human (or perhaps even *any*) life. So that leaves Mars. And while it's not exactly as welcoming as Mother Earth, Mars has a lot going for it as a potential colonization site.

Mars is our next-closest planet after Venus. It would take about 150 days, or five months, to travel there using existing spacecraft technology. Its proximity means that traveling to Mars and back, with a stopover someplace to conduct experiments and collect data, would be endurable for astronauts, even considering the toll that space travel takes on the human body. Various space agencies have already sent robotic missions to the planet—mostly one-way trips—so the technology has been at least partially tested.

Mars is also located within "the Goldilocks zone," a specific range of distance from the sun that corresponds to the range of temperatures at which liquid water can exist on the planet's surface. Also called the "habitable zone," it is the area that experts believe can support life—at least life as we recognize it on Earth. Technically, Mars does not always lie within this zone; it enters only when it nears perihelion (the point in its orbit when it is nearest the sun). It is close enough, however, that maintaining temperatures suitable for astronauts would be possible.

The atmosphere of Mars is very thin compared to Earth's, offering little protection from solar winds and radiation. The surface pressure on Mars is about 0.6 percent of that found on Earth, and its gravity is also lower, at only 37.5 percent of that on Earth; this could pose physical problems, such as bone loss, for astronauts.

The United States formally declared its commitment to Mars exploration on July 20, 1989, the 20th anniversary of the Apollo 11 moon landing, when then-President George H. W. Bush announced a long-term, multistage plan for NASA. The plan, which came to be known as the Space Exploration Initiative, was to start with the creation of a space station, followed by a return to the moon, where a research station was to be built, and conclude with a mission to Mars.

Although the Space Exploration Initiative is no longer reflective of NASA's plans, a trip to Mars is still in the works for the 2030s. Astronauts would be sent to orbit Mars and collect data before returning to Earth. This could be a prelude to a more permanent presence on the Red Planet.

H.G. WELLS
AND MARTIAN INVADERS

Herbert George (H.G.) Wells discovered his passion for science during childhood while recovering from a broken leg. Years later, another injury gave Wells ample time to peruse books, this time nudging him toward a career as a writer.

The War of the Worlds was Wells's fourth novel, serialized in 1897 and published as a book in 1898. Though well received, it was not considered his best work. It features a happy ending, but the plotline reflected the rising tensions among the world's economic powerhouses—a precursor to World War I.

The 1938 adaptation of the novel into a radio play by Orson Welles (pictured at left), presented as if an actual invasion was taking place while the radio audience tuned in, turned *The War of the Worlds* into a true sensation. Despite disclaimers that the tale was fictional, some listeners believed the play was a report of real events. The broadcast sent radio listeners into a panic, especially in New Jersey, where the Martians were supposed to have landed. Interestingly, many who believed the invasion was really happening thought the invaders were Nazis, not aliens, as the radio play was broadcast shortly before the onset of World War II. In the end, the radio play was so successful that H.G. Wells found himself pleased with the adaptation of his work—an idea that he had initially disliked.

THE GOLDEN AGE TO THE NEW WAVE

THE EXPLOSION OF SCIENCE FICTION AND THE RISE OF EXPERIMENTAL WORKS

Breakthroughs in science and science fiction tend to develop in tandem. Victorian writers, like Jules Verne, used what was known to science—and what was merely popular belief—to weave their tales of adventure and the future. H. G. Wells's *The War of the Worlds* was based on what astronomers knew about Mars at the time, and it extrapolated, as science fiction does, in logical ways so as to tell an engaging story. Without the early, blurry observations about the surface of Mars and, in particular, the work of astronomer Percival Lowell, who studied the planet through his telescope and made maps of features he thought he could make out, Wells might have chosen some other place of origin for his alien menace. But science said Mars might have once supported life—and maybe it still did. *The War of the Worlds* grew from there.

It wasn't until the '60s that the surface of Mars was observed in clearer detail, revealing a barren planet; between the Victorian era and the *Mariner* missions in the '60s and '70s, writers were free to imagine what shape a Martian civilization might take while still incorporating the advances of science. In 1912, Edgar Rice Burroughs, best known as the creator of *Tarzan*, published *A Princess of Mars*, in which he imagined a Mars that was less hostile than Wells's vision. Other writers, many of them penning pastiches and knockoffs of Wells's and Burroughs's work, found the popular "pulps"—digest-sized anthology magazines, essentially—a welcoming outlet for their myriad imaginings of the Red Planet.

THE GOLDEN AGE

The Golden Age of science fiction is generally agreed to have begun in 1938, the year that legendary editor John W. Campbell gained full editorial control over *Astounding Science-Fiction* magazine. It was also the same year C. S. Lewis, beloved author of the *Narnia* series, released *Out of the Silent Planet*, the first novel in his space trilogy. This was a deliberate (and theological) response to H. G. Wells, and though Mars was still a dying planet in Lewis's narrative, its

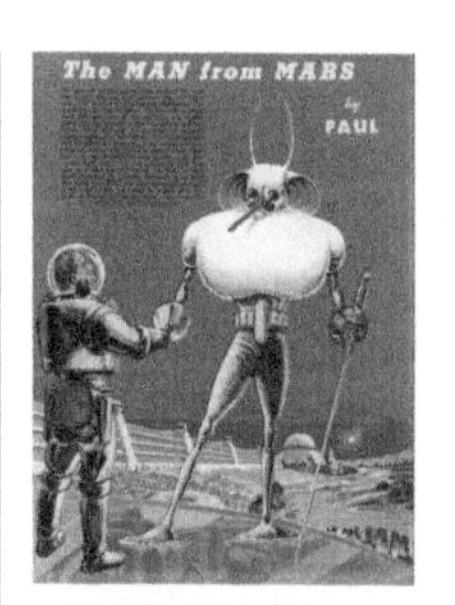

Martian life as sensationalized in popular films and novels of the 1950s.

intelligent inhabitants were resigned to extinction and chose not to invade Earth even though they had the technological means to do so.

Astounding Science-Fiction, originally titled *Astounding Stories*, was one of many pulp magazines to emerge in the '20s and '30s. Dominated by editor Hugo Gernsback, the Pulp Era was an age of gizmos and gadgets, in which the wow factor was more important than the characters. When Campbell took over *Astounding Science-Fiction*, he almost single-handedly ushered in the Golden Age, changing the focus of science fiction (Sci Fi) to characters and narratives while maintaining the genre's hard-science edge.

Some include the 1950s in their discussions of the Golden Age, and writer Robert Silverberg sees the '50s as the true Golden Age of Sci Fi. It was, at least, a transitional time, when the Sci Fi genre was coming into its own. The decade saw an explosion in science fiction and a number of books, now considered classics, that also happened to be set on Mars. Some of the now-beloved writers who wrote books about Mars in the '50s include Ray Bradbury, Lester del Rey, H. Beam Piper, Arthur C. Clarke, Isaac Asimov, John Wyndham, Kurt Vonnegut, and Harry Harrison. The list is a who's who of the founding fathers of science fiction (at the time, the world of Sci Fi authorship was primarily a boys' club, though more women joined the ranks each year).

Ray Bradbury's book *The Martian Chronicles* is a classic example of the genre. Previously published as a series of short stories, each piece was later strung together to create an episodic novel. In a way, the main character is Mars itself, and the book chronicles human attempts to settle the Red Planet as well as the indigenous Martians' efforts to keep out the so-called "Earth Men."

So many great Science Fiction books were written in the 1950s; it would take another book just to list and discuss them all—even if we kept the list to only those books featuring the Red Planet. If you're interested in digging in to these classic novels, check out: *Marooned on Mars* by Lester del Rey (1952), *Star of Ill-Omen* by Dennis Wheatley (1952), *No Man Friday* by Rex Gordon (1956), and *The Outward Urge* by John Wyndham (1959).

Films and books of the 1920s and '30s popularized Mars as a fantasy land of aliens and rocket ships.

CHANGE COMES

What was it about the late '40s and the '50s that made those years so welcoming to books about Mars in particular? Although it was dominated by Campbell's *Astounding*, the Golden Age saw the creation of a wide variety of new magazines publishing everything from the schlockiest adventure tales to the most serious of science-based stories. The adventurous spirit of the World War II era might have spurred reader desire for heroic tales of technology and science. But things were changing.

By the mid-'50s many magazines had folded, victims of an inflated market and declining readership. The introduction of the Comics Code in 1954 dealt a serious blow to comics by restricting the sort of content they could publish. By the late '50s, real technology was nearly as astounding as what writers had previously only dreamed of. Space travel had become a reality.

Robert Silverberg points out that magazines were not the only market for Sci Fi stories. While Sci Fi magazine sales declined throughout the decade until they were nearly gone by the late '50s, book publishers finally decided what to do with the genre. The venerable Sci Fi lines of publishers, Ballantine and Del Rey, began to publish both paperback and hardcover Sci Fi novels during this decade, giving the genre a gloss of respectability. The rush was over, but science fiction was here to stay.

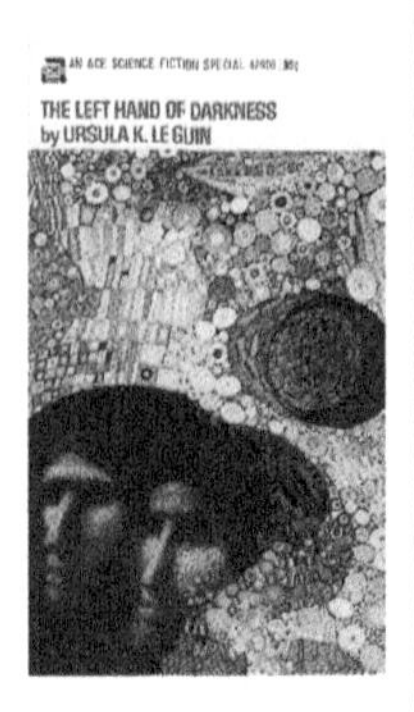

NOT JUST FOR MEN

In the 1960s, the New Wave of the Sci Fi genre included a number of important and influential women writers, such as Ursula K. Le Guin and Joanna Russ

NEW WAVE

If we think of the 1950s as the Sci Fi genre's teen years, the time when the genre grew quickly and caught a few glimpses of what it was to become, we can perhaps refer to the next era of Sci Fi as the genre's young-adult experimental phase—its college years, so to speak. Though the genre was maturing, it was hardly ready to settle down into one formulaic model. The '60s and '70s might not have had the excitement of the '50s, but Sci Fi published in those decades shows plenty of experimentation with the form and techniques of fiction itself.

Many of the New Wave writers viewed previous eras of science fiction as pulpy, juvenile, and badly written. They aspired to a more literary fiction and were less concerned with the actual science aspect of their stories. "Soft science fiction"—work using science as backdrop rather than focus, or that uses the soft sciences of anthropology, sociology, and the like—became more common than "hard science fiction"—work using science as its central topic, or that made use of the hard sciences of physics, engineering, and so on.

Despite the shift in focus for the genre and the fact that the surface of Mars was soon revealed to be barren rock where no intelligent life had ever lived, the Red Planet did not lose its appeal. In 1961, Robert Heinlein's *Stranger in a Strange Land* was published, with a premise not based on the science of Mars, but instead with a focus on how being raised on a different planet by an alien species might affect a human being.

SCIONS OF SPUTNIK

THE COLD WAR HEATS UP THE SPACE RACE

Space exploration was born from the Cold War posturing between two huge, powerful nations, both of which wanted to best the other, but neither of which wanted actual war to break out. The Cold War, lasting from the late 1940s to the early 1990s, was a period of fluctuating tensions and hostility between the US-led Western and Soviet-led Eastern Blocs. The tensions arose from the end of World War II and ideologically tore apart countries that had formerly been allies against Nazi Germany. The United States and the Soviet Union emerged as the two most powerful nations in the world. They were divided by vast economic and political differences, though no large-scale fighting ever broke out between them.

Both countries had nuclear arms and the capability to wipe out life on the planet, though neither side ultimately deployed this power. But the development of rockets that could be used for nuclear launches also benefited the space programs of both countries. Those same rockets could be used to send satellites, spacecraft, and even moon missions into space.

"Long live the world's first female cosmonaut" reads a Czech poster celebrating Valentina Tereshkova, who orbited the Earth in *Vostok 6*, in June 1963.

SPACE RACE

Instead of channeling the hostility between nations into outright aggression, the two countries' space programs gave them a safer, more productive arena. By competing to be the first—the first nation in space, the first to launch a living creature and bring it back safely, the first to send a human into orbit, the first to reach the moon—the superpowers could show who was best without coming to blows.

The race began when the United States announced its intent to launch satellites to celebrate the International Geophysical Year and the Soviets responded by also announcing a satellite program. In fact, the USSR got its satellite, *Sputnik 1*, into orbit first, with an October 1957 launch.

For many years, the cosmonauts of the Soviet Union and the astronauts of the United States were the only significant contributors to space exploration worldwide. Whenever the phrase "the space race" was mentioned, it was understood as a two-way race between the United States and the USSR in which the two kept pace and ran nearly parallel missions.

VOSTOK AND MERCURY: EARLY 1960s

The "winner" of the space race was by no means certain. The United States was first to launch an animal into space (in 1949, when it sent fruit flies and, later the same year, the first of several monkeys), but the USSR was the first to put one into orbit (a dog, in 1957). Yuri Gagarin, a Soviet cosmonaut, was both the first human in space and the first to orbit Earth when he launched into space while aboard the *Vostok 1* in April 1961. US astronaut Alan Shepard followed Gagarin into space only a month later, as the *Mercury* program debuted in May 1961; he became the first person to actually control the spacecraft himself, by using its rockets. Though Shepard didn't make it into orbit, John Glenn did in February 1962 during another Project Mercury mission.

During this era, the Soviets twice performed dual-spacecraft missions, in which two identical craft were launched on nearly identical orbits within 24 hours of each other (*Vostok 3* and *4*, and *Vostok 5* and *6*). They tested communication between each craft, though owing to slightly different orbits, they were not always close enough together. The USSR also sent the first woman into space as a propaganda move to gain one more "first" and demonstrate its superiority.

VOSKHOD AND GEMINI: MID-1960s

The USSR's Voskhod missions and the US Gemini program were both characterized by tests and further propaganda. The first rush of competition—to get people into space safely—was over, and though both nations aimed for greater things, they first had to find out how far they could push their technology. The US space program was also working toward a moon mission and used the Gemini missions as smaller-scale tests of the technology that would be needed to land astronauts on Earth's own natural satellite.

The Soviets were also planning their next phase of space exploration—what would become the Soyuz program—but it would not be ready to launch for some time. To test its technology and avoid looking lax next to the United States, Soviet scientists modified the remaining *Vostok* spacecraft for the Voskhod missions. *Voskhod 1* achieved two firsts in spaceflight: It was the first craft to have three crewmembers, and it was the first to test spaceflight without the use of space suits.

Though the USSR's Voskhod program got its start before the US Gemini program,

DID YOU KNOW

While Canada has yet to fly its own separate missions, the Canadian Space Agency has contributed both astronauts—including the much-loved Chris Hadfield—and equipment. The robotic arm used in shuttle missions since Columbia (STS-2) in 1981 was developed and built by the Canadian Space Agency and is appropriately called the Canadarm.

Artist Valentin Petrovich's 1963 poster proclaims "Our triumph in space is a hymn to the Soviet country!"

the Gemini missions accomplished some significant feats as well. On *Gemini 3*, the astronauts were the first to deliberately change their craft's orbit, and Gemini 5 set a mission-length record, with its astronauts spending nearly eight days in space—not coincidentally, long enough for an Earth to Moon voyage. *Gemini 6A* and *Gemini 7* were the first craft to perform a space rendezvous, matching orbits and maintaining distances as close as one foot.

Both countries also tested the spacewalk, sending cosmonauts and astronauts outside the safety of their spacecraft to perform tasks. Spacewalks were vital to get right. Merely floating in space in a space suit was dangerous, but if humans were going to go on the extended space flights necessary to explore the solar system, then they would need to be able to conduct repairs on their vehicles and capsules.

COOPERATION AND DERAILMENT

Realizing that two great nations could achieve a lot more together than they could in competition, US President John F. Kennedy proposed a joint US-Soviet space program. He announced the idea in a speech to the United Nations General Assembly on September 20, 1963. This was part of the United Nations' increasing involvement in space exploration, though Soviet Premier Nikita Khrushchev initially rejected the proposal. Khrushchev reportedly changed his mind later that year, but Kennedy's assassination on November 22, 1963 temporarily brought an end to hopes of a joint space program.

Kennedy's proposal was part of a series of debates between the United States and the USSR on the uses of space presented before the United Nations. Discussions began in 1958, resulting in the formation of the Committee on the Peaceful Uses of Outer Space the following year. The committee urged member nations to extend international law into space so that no one nation could claim any part of space for itself. Instead, the reasoning went, space exploration would be used for the good of all, much like scientific research in Antarctica. Both the United States and the USSR supported this ideal.

SOYUZ AND APOLLO: LATE 1960s TO EARLY 1970s

The next phase in space exploration, featuring the Apollo program in the United States and the Soyuz program in the Soviet Union, began with disaster for both nations. *Apollo 1* caught fire during a ground test, killing its three-member crew less than a month before its scheduled launch. The *Soyuz 1*, on the other hand, made it to space but was plagued by multiple technical problems that ultimately ended in the death of its sole crew-member, pilot Vladimir Komarov, during the craft's emergency reentry. This marked another first for the Soviets, albeit a sad one: Komarov's death was the first in-flight fatality in the history of space exploration.

After the initial disasters, both nations fixed the issues that had destroyed their missions, and they continued to develop their spacecraft. The USSR's Zond 4 mission completed a flight around the moon in 1968, and the Soyuz 4 and 5 missions completed the first docking of two manned spacecraft in 1969. That same year, *Apollo 8* was sent into lunar orbit, though it had to go without its lunar module, which wasn't ready in time to make the trip. The Apollo 8 mission was the first during which the *Saturn V* rocket was used, and its three-person crew was the first to leave low-Earth orbit for another celestial body. The crew's return was the US space program's first water landing of a spacecraft, though the Soviets had managed one the year before with the *Zond 5* (which carried several tortoises but no crew).

ONE SMALL STEP FOR ROBOTS SUCCESSFUL ROBOTIC (UN-CREWED) MISSIONS TO SPACE

1957

OCTOBER 4
Sputnik 1 (USSR)
First successful robotic mission for USSR, first artificial satellite around Earth

1958

JANUARY 31
Explorer 1 (USA)
First successful US robotic mission, first US satellite around Earth

1959

JANUARY 2
Luna 1 (USSR)
First lunar flyby

1960

Pioneer 5 (USA)
First solar monitor

The next several Apollo missions brought the US space program closer to a moon landing, as each mission tested more technology needed to land safely. *Apollo 9* took the lunar module for its first test flight in March 1969, while *Apollo 10* took the module into a close orbit of the moon, only 14.4 kilometers (8.9 miles) above the surface, the same point beyond which astronauts would need to begin a powered descent. That descent finally took place in July 1969 when the three astronauts of *Apollo 11* touched down after a three-day journey from Earth.

While the US space program continued making new strides, the Soviet program was in trouble. Launch failures and explosions prevented the USSR from sending a mission to the moon, and it lost significant ground in the space race as a result.

While the United States' moon landing in 1969 is seen as the most significant outcome of the space race, there was another milestone that often goes unrecognized. In 1972 the United States and the USSR agreed to a joint venture in space, which came to fruition three years later in the form of the Apollo-Soyuz Test Project.

SPACE STATIONS: SALYUT AND SKYLAB

Having lost the race to the moon, the Soviets decided to concentrate their energy on building space stations. They continued work on the Soyuz program, which would become instrumental in their space station efforts, and in April 1971 the USSR launched the first space station, Salyut 1. The initial attempt to dock a Soyuz craft with the station failed, but the next was successful and the Soyuz 11 crew stayed on Salyut 1 for a record 22 days. Unfortunately, the way home proved tragic—the crew asphyxiated due to a faulty cabin pressure valve.

After 175 days in orbit, *Salyut 1* burned up in Earth's atmosphere, despite efforts made to prevent such an event. The Soviets made five further attempts to launch Salyut-class stations, but none was successful.

Not content to remain complacent regarding space stations, the US space program launched *Skylab 1* in May 1973. The station was damaged during its ascent, but it was repaired

ONE SMALL STEP FOR ROBOTS SUCCESSFUL ROBOTIC (UN-CREWED) MISSIONS TO SPACE

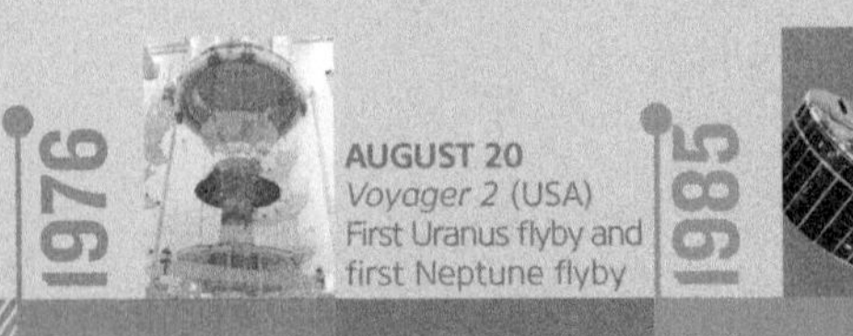

1976

JANUARY 16 *Helios 2* (USA and West Germany) First German successful robotic mission

1977

AUGUST 20 *Voyager 2* (USA) First Uranus flyby and first Neptune flyby

1985

JANUARY 7 *Sakigake* (Japan) First successful Japanese robotic mission

1985

JULY 2 *Giotto* (Europe) First successful European mission, first Comet Grigg-Skjellerup flyby

during subsequent missions and remained in use until its final mission in 1974, when its crew stayed in orbit for a new record of 84 days. The station finally reentered orbit and broke up in July 1979.

SHUTTLES, MIR, AND THE INTERNATIONAL SPACE STATION

In the West, we tend to think of the space shuttle program as an American affair, but it was at least partly an international effort. Other countries sometimes contributed technology and astronauts, but the US space program's biggest partner was its old rival, the USSR (and after the Soviet Union's dissolution, Russia).

The Soviets began construction of the space station Mir in 1986. It was the first modular space station, constructed in parts that were attached together after arriving in space. Mir was continuously inhabited by a changing crew of cosmonauts and remained in use for nearly 10 years.

1990
APRIL 25 Hubble Space Telescope (USA and Europe) First space telescope
1991
AUGUST 31 Yohkoh (Japan, USA, and United Kingdom) First successful UK robotic mission
1997
OCTOBER 15 Cassini-Huygens (USA and Europe) First Saturn orbiter and first probe of Saturn's moon Titan
2013
NOVEMBER 5 Mars Orbiter Mission (India) First successful Indian mission, first successful mission by an Asian country
isro

NASA's space shuttle program, officially known as the Space Transportation System, or STS (each mission was called STS but had a number appended to it for identification purposes), began in 1981 with a series of test flights. By the next year, the shuttles were flying the first of 135 missions launched between 1981 and 2011. The first shuttle, *Enterprise*, was a test vehicle with no orbital capabilities. It was followed by four fully operational vehicles *Columbia, Challenger, Discovery* and *Atlantis*. Of these, *Challenger* was lost during a launch failure in 1986 and *Columbia* was lost during a reentry failure in 1993. A fifth orbiter, *Endeavor*, was built in 1991 to replace *Challenger*.

The shuttles were flown on a variety of missions, including cooperative projects with Russia and the cosmonauts on Mir. During this time, cosmonauts flew on the shuttles, astronauts flew on board Russian Soyuz craft, and the shuttles docked numerous times with Mir. Finally, in 1998, the joint mission ended and Mir was intentionally deorbited in 2001.

The end of the shuttle program and Mir was not the end of joint space programs, however. On the contrary, space exploration only became even more international with the International Space Station (ISS). Construction began in 1999 with 10 pressurized modules that would become the core of the station.

MARTIANS ON SCREEN

1 **In 1953** the Looney Tunes cartoon spin-off *Duck Dodgers in the 24½th Century* debuted Marvin the Martian, a diminutive—almost cute—antagonist bent on destroying Earth because it was blocking his view of Venus.

2 **In 1962** a more menacing type of Martian appeared. "Mars Attacks" bubblegum cards, created by trading-card manufacturer Topps, told the story of a Mars on the edge of a destructive explosion and of Martians desperate to escape their home planet to take over and ravage Earth (and Earth's women, apparently). The 1996 Tim Burton movie based on the cards slightly changed the story but not the look of the aliens.

3 **The *My Favorite Martian* TV series,** which originally aired from 1963 to 1966, humanized Martians even more with its titular character, who appeared entirely human except for a pair of retractable antennas on his head.

FICTION OR FACT?

SCIENCE FICTION DRAWS ON FACTS, BUT SOMETIMES IT *BECOMES* FACT

CELL PHONES *STAR TREK* (1966)

The original *Star Trek* series had the *Enterprise*'s crewmembers talking to each other and to the ship using a device called a "communicator." The handheld, rectangular objects flipped open—much like the first flip-open cell phones that became widely available in the early 2000s.

TABLET COMPUTERS
ARTHUR C. CLARKE, *2001: A SPACE ODYSSEY* (1968)

In Clarke's book, people used thin tablet computers called newspads in everyday life to access the world's newspapers in electronic form. Though Clarke's newspad needed to be plugged in to access information, it closely resembles today's tablets in nearly every other way.

COMMUNICATIONS SATELLITES
ARTHUR C. CLARKE, *WIRELESS WORLD* (1945)

Clarke wrote a letter to the editor in the February 1945 issue of *Wireless World* magazine as well as an article in the October 1945 issue of the same magazine, discussing possible peacetime uses for the V-2 rockets the Allies had captured from Germany at the end of World War II, such as using the rockets to launch satellites into orbit. The type of orbit Clarke proposed, a geosynchronous orbit (an orbit harmonious with Earth's rotation) that would keep the satellite always over the same part of the surface, is still known as a "Clarke orbit."

EARBUDS RAY BRADBURY, *FAHRENHEIT 451* (1953)

In Bradbury's classic novel, people hold devices reminiscent of seashells by their ears to deliver audio entertainment only they can hear. At the time, such objects were unheard of, but in 1980 the first in-ear headphones—commonly known today as earbuds—were released for sale.

VIRTUAL REALITY RAY BRADBURY, "THE VELDT" (1951); ARTHUR C. CLARKE, *THE CITY AND THE STARS* (1956)

In Bradbury's short story, a family lives in a fully automated house, complete with a nursery that can read the minds of the family's children to create a real-seeming illusion of any place the children can imagine. Clarke's novel *The City and the Stars* takes virtual reality a step further with an imagining of virtual-reality video games (*Oculus Rift* and Sony's *Project Morpheus*, in fact, are nearly ready for commercial release) at a time when video games didn't yet exist.

SPACE AGE(NCIES)

EVERYONE WANTS TO BE AN ASTRONAUT (OR A COSMONAUT)

The list of national government space agencies is quite long. The biggest agencies are, naturally, those that have made the most significant contributions to space exploration—largely because getting to space is expensive. But smaller agencies, like Costa Rica's ACAE (Asociación Centroamericana de Aeronáutica y del Espacio) or the Belarus Space Agency, can make vital contributions—a sound reason for being included in international space projects. The agencies that follow are the biggest and, so far, the ones that have done the most to get humanity to the stars.

NASA NATIONAL AERONAUTICS AND SPACE ADMINISTRATION

COUNTRY: UNITED STATES OF AMERICA

NUMBER OF MARS MISSIONS: 13

LEARN MORE: NASA.GOV

The biggest and best-known space agency is NASA, which has been involved in space exploration since the very beginning. Although NASA was not the first agency to put a human in space, the agency was the first and only group to put humans on the moon. NASA has been the biggest player in Mars exploration, and it is the national agency most likely to be first to reach Mars with a manned mission. Of the Mars missions that NASA has launched, five are still transmitting data.

CNES CENTRE NATIONAL D'ÉTUDES SPATIALES

COUNTRY: FRANCE

NUMBER OF MARS MISSIONS: 0

LEARN MORE: CNES.FR

The French national space agency, CNES, was established in 1961. Though it once trained astronauts, that part of the agency's work was transferred to the European Space Agency in 2001. CNES continues to operate on smaller projects, contributing to the International Space Station and most recently working with Germany and other governments on a project to build a reusable launch vehicle by mid-2015.

ESA EUROPEAN SPACE AGENCY

COUNTRIES: EU MEMBER COUNTRIES, NORWAY, SWITZERLAND

NUMBER OF MARS MISSIONS: 2

LEARN MORE: ESA.INT/ESA

The European Space Agency is composed of the nations in the European Union, as well as Norway and Switzerland. It is close in size and significance to Russia's agency, and it has launched several important missions to study Earth's atmosphere, conduct astronomy, and send probes and orbiters out into the solar system. ESA's Mars Express orbiter program was only partially successful because its *Beagle 2* lander failed to communicate, but the orbiter part of the mission confirmed the presence of water ice at Mars's poles and continues to send back information. A new mission, called ExoMars, will include an orbiter and lander that will be followed by a rover. The European Space Agency includes several member countries—among them France, Germany, and Italy—that also have their own space agencies.

ISRO INDIAN SPACE RESEARCH ORGANIZATION

COUNTRY: INDIA

NUMBER OF MARS MISSIONS: 1

LEARN MORE: ISRO.GOV.IN

The ISRO is one of the largest government space agencies, though few people in Western countries know about it. It was established in 1969 and has developed and launched satellites and rockets. The ISRO sent a mission to the moon in 2008 and an orbiter to Mars in 2013, making India the first Asian country to successfully send a spacecraft to Mars.

DLR DEUTSCHES ZENTRUM FÜR LUFT-UND RAUMFAHRT

COUNTRY: GERMANY

NUMBER OF MARS MISSIONS: 2 (WITH THE ESA)

LEARN MORE: DLR.DE

The DLR is Germany's center for aerospace, energy, and transportation research, and it has conducted a wide range of projects both independently and with other nations. It was created in 1997 from a legacy of earlier institutes. Germany has had a number of astronauts fly aboard shuttle, Soyuz, and Mir missions.

JAXA JAPAN AEROSPACE EXPLORATION AGENCY

COUNTRY: JAPAN

NUMBER OF MARS MISSIONS: 1

LEARN MORE: GLOBAL.JAXA.JP

Japan's national space agency was born from a merger of three independent agencies in 2003. It is responsible for the development and launch of satellites and is also pursuing the exploration of asteroids and manned trips to the moon. One of JAXA's predecessor organizations sent an orbiter to Mars, but it was unable to achieve orbit. Other missions include astronomy, Earth observation, and solar observation.

ROSCOSMOS RUSSIAN FEDERAL SPACE AGENCY

COUNTRY: RUSSIA

NUMBER OF MARS MISSIONS: 4

LEARN MORE: EN.FEDERALSPACE.RU

During the Cold War, the USSR was the United States' biggest rival in space exploration. After the dissolution of the Soviet Union, it took time for the space program to recover, but Russia's Roscosmos is currently the second-largest national space agency. Russia has been instrumental in the construction of the International Space Station and landed several probes on Mars, but it has otherwise not been especially successful in exploring the Red Planet.

CSA CANADIAN SPACE AGENCY

COUNTRY: CANADA

NUMBER OF MARS MISSIONS: 0

LEARN MORE: ASC-CSA.GC.CA

The CSA has contributed important technology and highly trained astronauts to international space exploration efforts. Fourteen shuttle missions and two ISS missions have included Canadian astronauts. ISS Commander Chris Hadfield (Expedition 34/35) became an international celebrity for his social media posts of stunning photographs taken from the ISS and for performing David Bowie's "Space Oddity" in space.

ASI AGENZIA SPAZIALE ITALIANA

COUNTRY: ITALY

MARS MISSIONS: PROVIDED INSTRUMENTS FOR THE ESA'S *MARS EXPRESS* (2003) AND *ROSETTA* (2004), AS WELL AS RADAR FOR NASA'S *MARS RECONNAISSANCE ORBITER* (2005)

LEARN MORE: ASI.IT/EN

The ASI was established in 1988 to operate Italy's space exploration activities. The agency provides Italy's delegation to the European Space Agency and has worked on satellites, probes, and other spacecraft, both on its own and in collaboration with other countries.

CNSA CHINA NATIONAL SPACE ADMINISTRATION

COUNTRY: CHINA

NUMBER OF MARS MISSIONS: 1

LEARN MORE: CNSA.GOV.CN

Established in 1993, the China National Space Administration is continuing a Chinese space program that has been in place since 1956. The CNSA is cooperating with numerous other nations to develop space technology and explore the solar system. The *Yinghuo-1* orbiter, meant to study Mars, was lost when the Russian spacecraft carrying it failed and reentered Earth's atmosphere, where it burned up. Future plans include manned and unmanned missions to the moon and Mars.

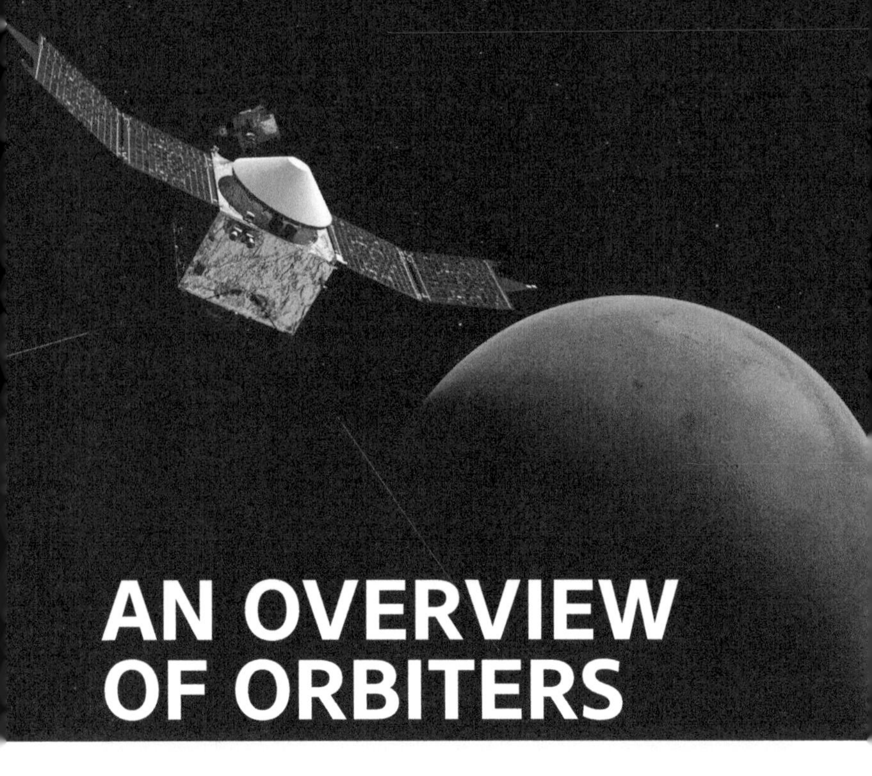

AN OVERVIEW OF ORBITERS

PEERING AT PLANETS FROM ABOVE

Launching an object into space is a difficult job; landing that same object safely is many times more difficult, especially if you're doing it from a distance with a significant communications lag. That's why the earliest successful

Above: This artist's concept depicts NASA's *Mars Atmosphere and Volatile Evolution* (MAVEN) spacecraft near Mars.

Mars missions were probes and flybys. The idea was to send a pair of craft: one to land on the surface and the other to remain in space to transmit data. At first, the craft remaining in space were on flyby missions, which were much easier to program than orbiting craft.

During the 1960s the Soviet Union sent up nine missions intended to reach Mars. All nine probes failed, some shortly after launch and others en route to the Red Planet. The first was launched in 1960, but only got 120 kilometers (74.5 miles) above Earth before reentering the atmosphere. The space race between the United States and USSR kept the Soviets going, and in 1971 their *Mars 3* lander touched down on Mars, accompanied by a space probe doing a flyby. Unfortunately, *Mars 3* ceased transmitting after only a few seconds.

Meanwhile, the United States was also sending missions intended to study Mars. In 1964 NASA's Jet Propulsion Laboratory sent *Mariner 3* and *Mariner 4* on flyby attempts. *Mariner 3* failed, but *Mariner 4* was successful, flying past Mars in July 1965 and sending back the first real close-up photographs of another planet. *Mariners* 6 and 7 reached Mars in 1969, and *Mariners* 8 and 9 were sent in 1971, though only 9 reached the planet. *Mariner 9* successfully entered Mars's orbit, becoming the first spacecraft to do so.

As mentioned above, the ideal robotic exploration of Mars would involve two crafts: one on the surface, collecting data, and a second craft in space (preferably in orbit) to take photographs and relay data from the landed craft back to Earth. To that end, in 1975 NASA launched two orbiters and two landers. Known as the Viking program, its aim was to look for signs of life and observe the Martian surface.

DID YOU KNOW

Whether a lengthy mission or a simple experiment, a failed space mission can be nearly as useful as a success. Every time equipment fails, designers learn more about its capabilities. And when an orbiter and a lander are sent together, useful information can still be obtained even if only one craft reaches its destination.

Once they worked out reliable ways to launch their crafts, the US and Soviet space programs discovered most of their problems were with landers, which were difficult to land safely on the surface and didn't always perform once there.

The next step for NASA was the launch of the *Mars Observer* in 1992, which failed but made way for the *Mars Global Surveyor*.

Mars Global Surveyor (MGS) launched in 1996 and reached Mars in 1997. Once it achieved a stable orbit, MGS began mapping the planet from a low-altitude, nearly polar orbit. Its primary mission—to map the entire surface of Mars—was completed in 2001, making it the first completely successful mission to Mars. *MGS* remained in orbit and continued to transmit data until it lost contact with Earth in 2006.

NASA followed *MGS* with a mission by *Mars Odyssey* in 2001, to look for evidence of water and volcanic activity. The craft was equipped with gamma ray and neutron spectrometers, and it found strong indications of water ice at the planet's south pole.

On November 12, 2014 the ESA's Philae lander successfully touched down on Comet 67P, achieving the first ever soft landing on a comet.

In 2003, the European Space Agency launched the *Mars Express*, an orbiter equipped with the *Beagle 2* lander. The lander entered Mars's atmosphere successfully, but the ESA was unable to maintain contact and the lander was declared lost in February 2004. The orbiter, however, still sent back plenty of useful information, including confirmation of the presence of water ice and carbon dioxide ice at Mars's south pole.

One of the most exciting missions—and definitely the most successful—was NASA's Mars Exploration Rover mission, which sent two rovers (*Spirit* and *Opportunity*) to the surface of Mars. The rovers were tasked with exploring the Martian surface, examining its rocks and soils. Both rovers continued to operate long after the conclusion of their 90-Martian-day mission.

The ESA has also maintained interest in Mars, sending its *Rosetta* space probe to slingshot around and observe Mars, though its primary target was a comet. Russia's Roscosmos, too, is still in the space game, though its Phobos-Grunt mission suffered a complete failure in 2012. The mission had two objectives: to drop a lander, which was intended to retrieve a sample from the surface of Mars's moon, Phobos; and to deliver China's Yinghuo-1 satellite to Mars's orbit.

NASA launched another very successful rover in 2011 with the Mars Science Laboratory mission, which carried the rover *Curiosity*. It also maintains an interest in orbiters alone and launched *Maven* in 2013 to study the Martian atmosphere. It reached Mars in 2014 and has been sending back information since.

In 2013, India joined in the scientific study of Mars with the launch of its *Mars Orbiter Mission*, or *Mangalyaan*. It reached Mars's orbit in September 2014 to study the Martian atmosphere, though its primary mission was a demonstration of technology.

With this program, India became the fourth nation to successfully send a mission to Mars and the first to achieve orbit on its first attempt.

NASA's most recent Mars orbiter mission is the *Mars Reconnaissance Orbiter* (or *MRO*), launched in August 2005 and falling into Mars's orbit in November 2006. The *MRO* carries cameras, spectrometers, radar, and other instruments for analyzing the planet's surface and monitoring its atmospheric conditions. In addition to its valuable contributions to Mars science in general, the *MRO* will be instrumental in selecting landing sites for future missions, observing landing conditions, and hosting a communications system.

THE TOP 10 OUTER SPACE FIRSTS

1 First long-range rocket (1942) During World War II Nazi Germany developed the first rockets to bomb Allied targets without having to fly over them in airplanes. After the war, the Allies used V-2 rockets to test the effects of sending a living creature into space.

2 First animal orbits Earth (1957) Soviet researchers launched a dog named Laika into space aboard *Sputnik 2*. Unfortunately, she died from overheating within a few hours.

3 First animals to orbit Earth and return alive (1960) Soviet dogs Belka and Strelka were launched with *Sputnik 5* and spent a day aboard before retuning to Earth alive. Also on board and returning safely were a rabbit, 2 rats, 42 mice, and some plants and fungi.

4 First human in space (1961) Soviet cosmonaut Yuri Gagarin was the first human in space and the first human to successfully orbit Earth, when he rode the *Vostok 1* space capsule beyond the border between the atmosphere and space, and around the planet.

5 First American in space (1961) Alan Shepard (right) became the first American in space aboard *Mercury 7* but was not the first to orbit Earth. That honor went to John Glenn in 1962, during the Mercury-Atlas 6 mission.

6 First unmanned mission to Mars (1964) Launched in 1964, the *US Mariner 4* probe made a flyby of Mars in July the following year. The photographs it sent back were another milestone: they provided our first up-close look at the surface of another planet.

7 First spacewalk (1965) Cosmonaut Alexei Leonov was the first human to "spacewalk" when he spent 12 minutes outside his craft during the Voskhod 2 mission.

8 First humans on the moon (1969) Neil Armstrong and Buzz Aldrin landed on the moon during the Apollo 11 mission.

9 First spacecraft to achieve a soft landing on Mars

(1971) The Soviet Mars probe program launched several probe and orbiter missions during the 1960s and '70s, and in 1971 it finally had some success. The *Mars 3* lander made it to the surface but unfortunately failed after only 14.5 seconds of transmitting.

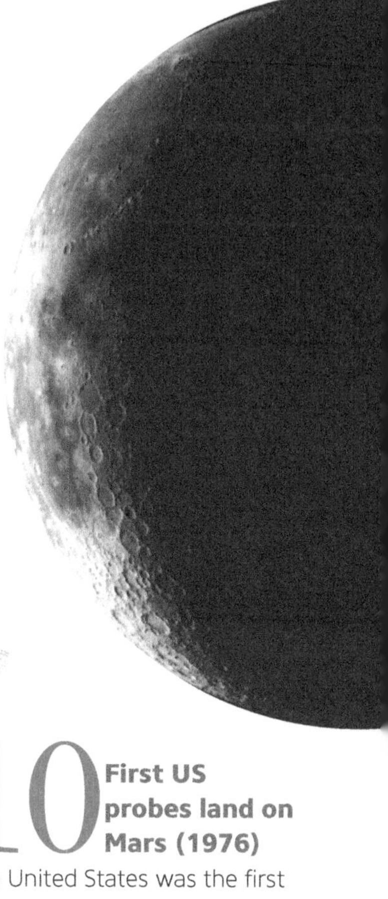

10 First US probes land on Mars (1976)

The United States was the first nation to successfully land equipment on Mars when the twin *Viking* landers arrived on the planet's surface.

BEYOND BOWIE

SOUNDTRACKS FOR THE RED PLANET

Just as Mars has been an inspiration for writers since Victorian times, it has also been an inspiration for musicians. One of the earliest Mars-inspired musical pieces was Gustav Holst's "Mars," one of seven parts in his *Planets* suite. Each movement is named for one of the planets, though Earth is not included because Holst's inspiration was more astrological than astronomical (and Pluto had not been discovered—or demoted from planet status—yet). Since then, there have been songs in many musical styles—from blues to pop, to punk, to grunge rock, and more—to suit any musical taste.

1918 GUSTAV HOLST
Mars, The Bringer of War

1968 SUN RA
Blues on Planet Mars

1971 DAVID BOWIE
Life on Mars?

1972 T. REX
Ballrooms of Mars

1975 WINGS
Venus and Mars

1976 DEXTER WANSEL
Life on Mars

1978 THE MISFITS
Teenagers from Mars

1978 JEFF WAYNE
Jeff Wayne's Musical Version of The War of the Worlds

1991 PIXIES
Bird Dream of the Olympus Mons

1993 THE MARTIAN
Journey to the Polar Cap

1996 ASH
Girl from Mars

1999 KELIS
Mars

2000 ENNIO MORRICONE
A Martian

2001 THE FLAMING LIPS
Approaching Pavonis Mons by Balloon

2004 MOUSE ON MARS
Send Me Shivers

2010 SYMPHONY OF SCIENCE
The Case for Mars

2011 COLDPLAY
Moving to Mars

RED ROVER, RED ROVERS

HOW EXTRATERRESTRIAL OFF-ROADERS HELP US UNDERSTAND THE MARTIAN LANDSCAPE

After flyby craft and orbiters, the logical progression in our quest to explore other planets is to send landers and rovers.

After a history of failed landers as the USSR attempted to get its Mars landers to the surface in the 1960s and '70s, NASA had success with the two *Viking* landers in 1975. Russia continued the USSR's streak of bad luck with two failed *Phobos* landers in 2012, while NASA has gone on to further successes. Other countries have joined in on space exploration, turning what was once a rivalry between two superpowers into an increasingly international scientific effort, especially as other nations contribute knowhow and technology to each other's missions.

PATHFINDER AND *SOJOURNER*

NASA had a number of early successes landing probes on Mars; the next logical step was to design a rover that could move around the surface, traveling to particular targets rather than simply exploring its surroundings wherever it happened to land. (Though of course all of the mission landing sites were chosen with care, hitting the exact target was not always possible.) The Pathfinder mission launched in 1997, and its rover *Sojourner* landed on Mars without incident in July that year. The tiny rover, also known as the *Microrover*, weighed a mere 11.5 kilograms on Earth (about 4.5 kilograms on the surface of Mars), and had only 64 kilobytes of RAM. *Sojourner* was designed for a project lasting only seven Martian days, though it was hoped the mission might be extended to 30 sols, the term scientists use to define the period of a day on Mars. Sojourner vastly exceeded expectations by

Left: Soil clinging to the wheels of NASA's Mars rover *Curiosity* can be seen in this image taken by the rover's Navigation Camera.

remaining active for 83 Martian days, traveling about 100 meters and remaining in contact with Earth until September 27, 1997.

Sojourner carried two black-and-white cameras and one color camera, as well as an APXS (Alpha Proton X-ray Spectrometer) very similar to the one carried by the failed *Mars 96* lander. The APXS was capable of examining the elemental composition of the soil and rocks on Mars.

MARS ODYSSEY

Originally part of NASA's Mars Surveyor program of 2001 and intended to be accompanied by a lander (the lander was canceled the year before), *Odyssey* launched on a *Delta II* rocket in April 2001. The orbiter's mission is to look for the presence of water and ice, or at least evidence of them in Mars's past, and to study the planet's geology and radiation to help NASA scientists determine the risks that a manned mission would face.

Odyssey is equipped with a thermal imager and spectrometers, and it is in a polar orbit at about 3,800 kilometers above the Martian surface. This orbiter has been circling around Mars for a record-setting 13 years; it's the longest-serving spacecraft studying the Red Planet. It continues to pursue its primary mission while also acting as a communications relay for subsequent missions. *Odyssey* is significant to the history of rovers on Mars because it delivered communications to Earth from the MER rovers *Spirit* and *Opportunity*, the MSL rover *Curiosity*, and the *Phoenix* lander (it also continues to relay information from the still-active *Opportunity* and *Curiosity*).

MARS EXPRESS AND THE *BEAGLE 2*

The European Space Agency launched the *Mars Express* orbiter in June 2003. Its primary mission, to look for evidence of water and carbon dioxide ice on Mars, was very successful. While it completed its mission in November 2005, the orbiter continues to observe and study the Red Planet. *Mars Express* had a second mission that did not fare so well. It dropped the *Beagle 2* lander as scheduled, but the fate of the lander remained unknown for many years. The ground crew lost contact as it descended to the surface, and contact was not reestablished. It was assumed that *Beagle 2*'s parachutes malfunctioned and that the lander burned up or crashed. Images from NASA's *Mars Reconnaissance Orbiter* showed in January 2015 that the *Beagle 2* had landed intact, but some of its solar panels had failed to deploy and blocked the lander's communications antenna. If it had been successful, the *Beagle 2* would have searched for life on Mars by examining its landing site's geology, chemistry, physical properties, and other information.

The lander missions of many nations have experienced both failures and successes, but so far every rover mission has done what it was supposed to—sometimes a lot more than it was designed for. NASA put four rovers on Mars—*Sojourner, Spirit, Opportunity, and Curiosity*—and all four did their jobs. *Two, Opportunity* and *Curiosity,* continue to travel to the Red Planet and send back new discoveries.

[**The *Beagle 2* rover** was named after the British ship *HMS Beagle*, which was the vessel Charles Darwin voyaged on in the 1830s, during which time he collected the specimens that would laterhelp him formulate the theory of evolution by natural selection. Finding signs of Martian life would have been the perfect achievement for the *Beagle 2*, so the failure of the lander is doubly unfortunate.]

MER: MARS EXPLORATION ROVERS

In early 2004, the twin MER rovers landed on Mars about a month apart and on opposite sides of the planet. Both rovers operated as hoped, driving over the surface of the Red Planet, investigating the targets chosen for them, and looking for signs of water in Mars's past. They completed their 90-Martian-day missions in March 2004, making the operation a success.

To the excitement of many, once the MER primary missions were over, the rovers remained operational, and both had their missions extended. For the next several years, *Spirit* and *Opportunity* continued to rove Mars, traveling to different features of interest, including rocks, ridges, dunes, and hills. *Spirit* finally got trapped in sand on April 29, 2009, and engineers were unable to free it. All was not lost, however, as *Spirit* continued to operate as a stationary science lab, investigating the sand in which it was trapped in more detail and continuing to send back photographs of the nearby surface. On May 25, 2011, after a lengthy hiatus in communications, *Spirit*'s mission finally came to an end. In 2013, scientists examining samples collected by *Spirit* earlier in its mission found that the 3.7-billion-year-old rocks contained fives times as much nickel as the 180-million-to 1.4-billion-year-old Martian meteorites that had been collected on Earth. This suggests that the rocks were formed in an oxygen-rich environment, but the meteorites were not. In other words, Mars once had more oxygen in its atmosphere.

Opportunity fared better than its twin and continues to explore the surface of Mars, visiting more sites of interest and continuing to operate long past the 10th anniversary of its landing. In 2013, the rover discovered Martian rocks containing clay, a good indication that Mars contains—or once contained—neutral-pH water, in turn an indication that life was possible. *Opportunity* alone has traveled more than 30 kilometers on the surface of Mars.

PHOENIX

Launched by NASA on August 4, 2007, the *Phoenix* lander touched down on Mars on May 25, 2008. It dug into the Martian soil and confirmed the presence of water ice beneath the planet's surface. Unfortunately, the harsh Mars winter damaged *Phoenix*'s solar panels and NASA was unable to reestablish communication with the lander after November 2008.

MARS SCIENCE LABORATORY

In November 2011 NASA launched an *Atlas* 5 rocket toward Mars, carrying the Mars Science Laboratory mission. The rover *Curiosity* is considerably larger than previous rovers—too large to land with airbags the way *Spirit* and *Opportunity* did. Instead, *Curiosity* was equipped with a rocket-powered sky crane to lower it to the surface. The landing, referred to by the NASA crew as "seven minutes of terror," took place successfully on August 6, 2012, at Gale Crater, near the Red Planet's equator. *Curiosity* almost immediately shot and sent back a panoramic photograph of its surroundings. The rover's mission, like those before it, was to look for possible signs of life and for the conditions necessary to have supported life in the past. In January 2013, the rover drilled into the Martian rock and discovered that rocks below the surface contained clay as well as minerals that had been in contact with water. This conclusion was supported by analysis of rocks collectedby the earlier *Spirit* and *Opportunity* rovers. When NASA scientists examined the samples, they found clay minerals, sulfates, and other chemicals that would most likely have formed in the presence of neutral-pH water that might have been able to support life.

Referred to by the NASA crew as "seven minutes of terror," the Curiosity *landing took place successfully near the Red Planet's equator.*

Mars scientists had long known of the presence of what appeared to be streambeds formed by running water, and *Curiosity* got up close to one such streambed in May 2013, using information gleaned from the mapping and photographs provided by earlier flybys and orbiters. The rocks in the ancient bed had rounded edges, indicating they had probably been "polished" by tumbling around in running water, just like rocks on Earth today.

In addition to studying the surface of Mars, *Curiosity* collects data on the atmosphere, and evidence it collected in 2013 suggests that a catastrophic event might have been responsible for destroying the early Martian atmosphere.

Although both *Opportunity* and *Curiosity* continue to operate, there is a lot more of Mars to explore, and neither NASA nor the other space agencies have given up on robotic vehicles—even with the ever-increasing likelihood of a manned mission to the Red Planet.

The European Space Agency and Russia's Roscosmos plan to launch the ExoMars mission in the coming years with an orbiter, followed by a rover in 2018. The Indian Space Research Organization will follow up its own Mars Orbiter mission with a project planned for 2018-2020, which will include a lander and a rover. China also plans a rover mission to Mars in 2020, as does NASA (which will include instruments contributed by France, Norway, and Spain), so there should be plenty of exciting finds in the years between now and when humans finally land on the planet.

THE VIEW FROM MARS

STANDING ON THE RED PLANET, LOOKING BACK AT THE BLUE DOT

Astronomy on Mars isn't all that different from astronomy on Earth, save for one thing: from Mars, our own home planet is visible. From Mars, the evening and morning "star" isn't Venus, like it is on Earth. Instead, Mars's evening star is Earth itself, accompanied by a smaller bright dot, our moon. And just like Venus and Mars can be seen as discs that go through phases (in the same way the moon waxes and wanes) through a backyard telescope on Earth, a Martian hobby astronomer—if there ever is such a thing—would see Earth as a disc that goes through a regular cycle of growing and diminishing crescent shapes.

The rest of the planets in the solar system would be similar to what we see from Earth, except they'd be smaller or larger, as Mars is farther away from or closer to them. And the stars would be so close to identical that it makes no real matter, because on the scale of the universe, Earth and Mars are not far enough apart to make a difference in the night sky's appearance. Because of the lower atmospheric pressure on Mars as compared with Earth, the stars would not appear to twinkle, and

Mars's pole star (the star around which all the others appear to rotate) would be Deneb in the constellation Cygnus instead of Earth's Polaris.

If astronomy on Mars is so similar to what we would see on Earth, why should we care about it? Aside from Mars's frequent dust storms and occasional ice clouds, the planet's thin atmosphere—which is mostly transparent to UV light due to its lack of an ozone layer—makes it a better location than Earth from which to conduct UV astronomy.

UV, or ultraviolet, astronomy is the observation of ultraviolet wavelengths and can be used to study chemical composition, density, and temperature of objects in the universe. In particular, UV astronomy is good for observing objects too hot to be emitting much radiation in the cooler, visible-light spectrum.

Mars is a good place to do astronomy for some of the same reasons the Hubble Space Telescope takes such excellent pictures. It's outside of Earth's atmosphere, free from all the distortions caused by our thick, protective blanket of warm gases.

To be completely free of distortion, of course, the best place to be is right out in space. But floating in space has its issues, too. Space-based equipment is a lot more difficult to access and repair than Earth-based equipment. If there were a colony or research station on Mars, then Mars-based equipment would also be easier to access, repair, and adjust than space telescopes.

MAPPING MODERN MARS

THE earliest features to be mapped on Mars were the dark and light areas, known today as "albedo features" ("albedo" refers to how much light is reflected from a surface). When the first spacecraft reached Mars in the 1960s, it became obvious that what early observers had assumed were seas, plains, or patches of vegetation were actually just the different colors of the planet's barren surface.

Victorian astronomers and even 20th-century observers who predated the 1975 Viking mission assigned romantic names to light and dark regions on Mars, like *Mare Australe* ("Southern Sea"), *Aetheria*, *Argyre*, and *Xanthe*, none of which make sense any longer. These astronomers mapped a place that existed mostly in the imagination (see map on the opposite page, dated 1890). But when they renamed Mars's landscape with accuracy, modern scientists retained earlier labels as often as they could.

For example, *Argyre*, a pale area below the dark patch formerly known as "Mare Erythraeum," roughly corresponds to what is now known as *Argyre Planitia*, a huge impact basin that might have once been filled with liquid water. In 1879, Italian astronomer Giovanni Schiaparelli saw a bright spot on the surface and named it *Nix Olympica*, or "Snows of Olympus." When it was discovered that this spot was actually an immense extinct volcano, scientists kept part of the old name by calling it *Olympus Mons* (Mount Olympus).

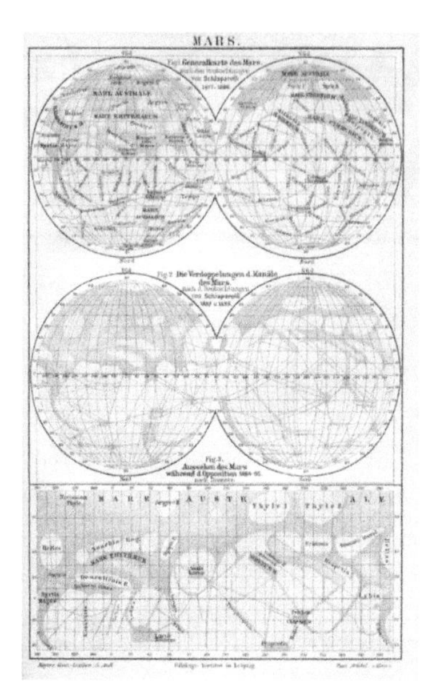

In other cases, the old names no longer fit at all. Dark and light areas have sometimes shifted so much that the old maps don't match current ones in any way. Large, general areas were therefore named for their physical features, with words like *terra* ("land"), *planitia* ("plain"), or *valles* ("valley") appended to what seem to be the best-fitting names from the old maps.

Other features, like individual craters and valleys that could not have been observed until at least the 1960s, are named according to a specific scheme. Large craters at least 60 kilometers across are named for deceased scientists and writers, while smaller craters are named after small towns and villages across the world. Large valleys are named for "Mars" or "star" in assorted languages, and small valleys are named for rivers.

WHERE'S THE WATER?

LOOKING FOR LIQUID

Except for a brief period of disillusionment following the Viking discoveries that Mars has a barren, dusty, wind-swept surface, scientists have nearly always thought that Mars had—or used to have—water. In the very earliest days of Mars astronomy, when Victorian scholars were peering at distance-blurred images in their basic telescopes, the dark splotches and streaks were thought to be bodies of water. By the time Percival Lowell began making his systematic observations in the 1890s, the thinking had shifted. Actual rivers and canals would have been too narrow to see from so far away, so Lowell proposed that the dark streaks were bands of vegetation growing alongside waterways—but that water was still definitely there.

The 1960s, of course, brought the knowledge that there were no watercourses and vegetation on Mars. But it wasn't long before evidence began to build that while Mars might be dry

In 1996, NASA scientist David S. McKay announced that his research team, based in Houston, had identified what might be fossil microbes in the Martian meteorite officially known as ALH84001. The findings were extremely controversial and not definitive either way.

(or rather, that it might *appear* to be dry) today, it might have been much wetter in the past. Mars might even have been a world of seas and rivers and life—perhaps not quite like early science fiction authors had dreamed, but exciting nonetheless.

One early indication that the Red Planet might be more than it appeared was found in the seemingly mundane discovery that the basalt rock comprising much of the planet came in two distinct varieties, one of which was much more glassy than the other, known to geologists as basaltic andesite (basalt). This type of rock is found closer to the poles than the equator of Mars, and on Earth it forms when lava comes into contact with water or ice, which make it cool much more quickly than it would otherwise.

Basalt is not the only indication that Mars might once have had ice or water beneath the surface. Some areas exhibit patterning that on Earth is characteristic of places with permafrost. As the upper layers of frost thaw and refreeze, they can break the surface into what is called "patterned ground." Also, the more northerly and southerly parts of the planet have odd craters surrounded by what look like landforms of heavy

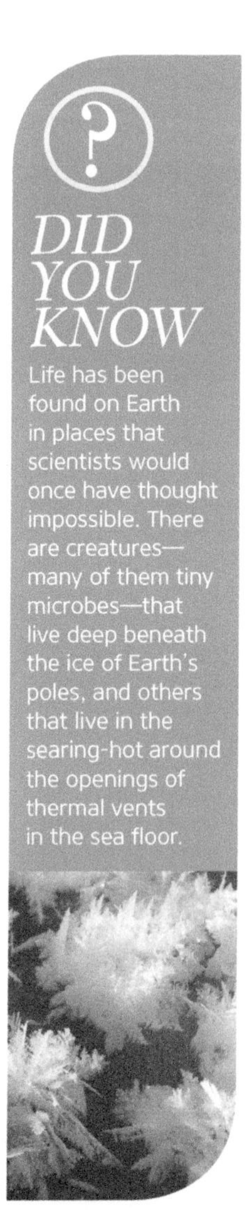

DID YOU KNOW

Life has been found on Earth in places that scientists would once have thought impossible. There are creatures—many of them tiny microbes—that live deep beneath the ice of Earth's poles, and others that live in the searing-hot around the openings of thermal vents in the sea floor.

VALLEYS/ VALLYS

Although it is frequently compared to Earth's Grand Canyon, Valles Marineris on Mars was not created by water erosion. Instead, it was formed primarily along geological faults. Flowing water almost certainly played a role in shaping and enlarging the Valles once the original fissures opened up, however.

mud, which is probably exactly what they are. When meteorites hit the planet and formed a crater, the ice underneath ground surfaces would have melted from the impact, causing the resulting craters to eject blankets of muddy soil and rock that froze around the crater. These craters are usually shallower than the more well-defined type, because that same mud would have slumped back into the impact area and filled it in partway.

It was known that Mars had polar caps quite early on—even the Victorians noted the bright southern cap and speculated it might be like Earth's poles—but no one could know for sure that the ice was water until orbiters carrying the appropriate equipment could test them. The physical evidence, therefore, was important in the early days, so when it was confirmed that Mars has water ice (as well as carbon dioxide or "dry ice") at its poles it was not surprising. The various lines of evidence also showed that it wasn't just the poles that once had water, but that it had extended much closer to the equator than today's ice caps.

Craters and basalts provide suggestive evidence rather than definitive proof. There is certainly water ice at Mars's poles, and this *might* have once extended much closer to the equator. With each new mission, though, the evidence is increasing. There are minerals in the soils of Mars—hematite and goethite, for example—that form in association with water on Earth, and it is reasonable to suppose they formed the same way on Mars, too. And the closer scientists get with better cameras and rover visits, the more certain it seems that what appear to be water-polished pebbles, outflow channels and dry riverbeds, canyons, and water-borne sediment deposits are the real thing.

In March 2015, NASA scientists presented evidence for an ancient Martian ocean in the planet's northern hemisphere. They measured the levels of water and deuterium in Mars's atmosphere and compared them to those on Earth. The conclusion—not yet universally accepted among those who study Mars—was that ancient Mars had much higher amounts of water. If Mars did have more water, it must also have been warmer and had a denser atmosphere—points that do not yet have significant evidence.

But assuming Mars was once a wetter place, and perhaps a warmer one with a thicker, more oxygen-rich atmosphere, what sort of life might it have supported? And could there still be life there today?

If there is life to be found on Mars, especially if that life still thrives today, it is likely to be microbial—i.e., tiny and hardy. *Curiosity* found evidence of organic carbon in Gale Crater in 2013, but it is unknown whether the carbon resulted from living organisms or from the cosmic dust that continually deposits organics on the planet.

Curiosity's mission has demonstrated, to the satisfaction of most scientists, that early Mars was habitable, at least by organisms that could have derived energy from the inorganic substances in their environments. Called autotrophic organisms, these microbes produce complex organic compounds from sunlight or chemical reactions. Chemotrophic organisms, similarly, produce energy through oxidation of either organic or inorganic substances, while chemolithotrophic organisms can convert carbon dioxide to glucose. These are the organisms found at the very base of the food chain, and they might be some of the earliest forms of life.

Knowing what types of life are the most likely to be found on Mars enables scientists to better understand what they should have the rovers search for. By the time the ESA-Roscosmos ExoMars mission reaches the Red Planet, scientists might know exactly what to search for and where would be best to look for them.

ARE MARTIANS ALREADY AMONG US?

WE WANT TO BELIEVE WE ARE NOT ALONE

IT seems that for as long as we have known the universe's vastness, humanity has wondered if we share the cosmos with other beings. And the larger the universe is, the more insignificant human beings seem, and the lonelier we feel.

In 1968 a Swiss hotel manager, Erich von Däniken, wrote a slender book titled *Chariots of the Gods? Unsolved Mysteries of the Past*. In the book, Däniken examined artifacts from various cultures all over the world, while incorporating the possibility that alien beings may have visited Earth in the past. He found sculptures he thought looked like airplanes, rock art that appeared to resemble tall, bipedal creatures wearing bubble-like helmets, and a clay pot with metal fragments he thought were the remains of a battery, the likes of which would not be discovered by humans for centuries after its historical origin.

Taken out of cultural context and looked at as a whole, Däniken's "evidence" seemed utterly amazing. But place each item into the culture from which it came, and most of them were much more mundane (or were amazing in purely human ways). Since then, other writers have taken up Däniken's cause, often seeing themselves as rebels against science and academia, which can seem adverse to change.

But supposing the "ancient aliens" hypothesis is bunk, what about more recent alien encounters? Quite a few people claim to have been abducted by aliens, there have been thousands of reported UFO sightings, and there really is a not-so-secret military base at Area 51 in the Nevada desert. Might the US government be hiding evidence of aliens?

DID YOU KNOW

In Vedic mythology, the universe was hatched from a cosmic egg that came into being out of nothingness. The ancient Sanskrit speakers knew nothing of modern astronomy or cosmology, but from our perspective far in their future, it's not too hard to imagine the Big Bang as a cosmic hatching from nothing.

In 1947 a mysterious craft crash-landed near Roswell, New Mexico. At the time the Air Force claimed the downed craft was a simple weather balloon, which was accepted by most people until the 1970s, when interest in UFOs led some to look more closely. The Air Force was, indeed, hiding something, but it wasn't a UFO. The craft was a balloon—a Project Mogul balloon used in nuclear test monitoring, not a weather balloon. The Air Force hadn't wanted to alarm anyone by revealing the balloon's actual purpose.

While it is possible that intelligent alien life has visited Earth, of course, it may not be as likely as some people would like us to believe. Although the possibility of alien life is astronomically high (probably—it is also possible to calculate the chances as just about exactly equal to none, depending on how you think about the universe), the likelihood of *intelligent* life may be much smaller, and the possibility of such life finding its way to us, or even wanting to find its way to us, is vanishingly small.

9
GREGORY

HOW DO WE BUILD IT?

DESIGNING FOR SPACE PRESENTS UNIQUE PROBLEMS

Aside from constructing a suitable spacecraft, there are still a lot of things to consider when planning a colony or even just a quick jaunt to another planet.

Everything sent into space has to be as light as possible. It's expensive and difficult to launch materials, and every gram counts. Once in space, weight is less of a concern, as it is mostly momentum that gets a spacecraft across long distances, but escaping the force of Earth's gravity is another thing entirely, as is landing on Mars. Volume is an issue, too; there isn't much room in a spacecraft, so everything on board has to be small or collapsible.

Then, assuming everything arrives on Mars intact, it has to be quick and easy to set up, and designed to give astronauts as much comfort as possible in a small space, on an essentially barren world. A Mars colony has to be completely self-sufficient, using only the materials brought along, and those available locally on Mars. It might be possible to send more supplies from Earth, but counting on that possibility means risking lengthy and potentially deadly delays.

Since Mars is too far away to test equipment to make sure it's going to work, how do scientists and engineers design for the

Red Planet? Fortunately, there are ways to simulate many of the situations astronauts will encounter, and there are environments here on Earth that can stand in for Mars—at least as far as basic testing goes.

TESTING WEIGHTLESSNESS

Since the earliest days of space exploration, it has been necessary for scientists to test the effects of zero gravity. It is essential for astronauts to practice living in and working with weightlessness.

Working in a weightless environment has unique challenges. Tools may drift away if they are set down, and gravity cannot be relied on for leverage. To help astronauts get used to these conditions, NASA and other space agencies use big swimming pools. The equipment astronauts will work with is sunk in the bottom of a deep pool, and the crew don their spacesuits to enter the water. It's not an exact simulation, but it mirrors conditions well enough to prepare crew for the realities of working in space. And because the astronauts can wear virtually identical suits in the pool as in space, they can practice moving in their cumbersome protective gear, too.

Another way to test endurance in a weightless environment is parabolic flights on large airplanes. The swooping, wildly oscillating flights can only simulate zero gravity for about 20 seconds at a time, but it's enough to give astronauts a taste and to test the functioning of equipment.

SIMULATED LIVING CONDITIONS

Because human psychology is as much an issue for living in space as the environment itself, there have been assorted attempts to simulate the kind of isolation colonists would face. To determine what personality characteristics are best and worst for selecting colonists, researchers observe people limited to strict rules, a controlled diet, and a single small space with few people for company.

One such experiment, Russia's Martian Surface Simulator, subjects candidates to an assortment of psychological tests to determine and prevent the negative effects a space colony might have on the human mind. These negative effects can include depression, aggression, sexual assault, and cultural misunderstandings (among an international crew). Without addressing these issues, a Mars voyage could be doomed from the start, no matter how good the technology supporting it is.

SIMULATED ENVIRONMENT

The most obvious challenge facing a Mars exploration crew or colony is, of course, Mars itself. It's a cold, barren, and dusty planet; it has little atmosphere; and it has less surface gravity than Earth. Not all of these conditions can be simulated at once, but there are places on our own

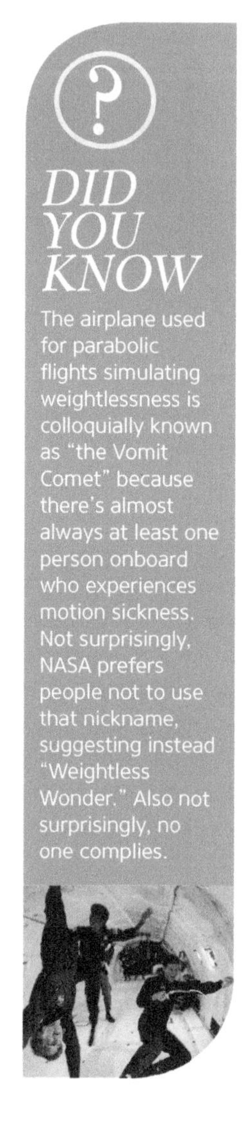

? DID YOU KNOW

The airplane used for parabolic flights simulating weightlessness is colloquially known as "the Vomit Comet" because there's almost always at least one person onboard who experiences motion sickness. Not surprisingly, NASA prefers people not to use that nickname, suggesting instead "Weightless Wonder." Also not surprisingly, no one complies.

planet similar to those on Mars, and they can be used to develop the technology that will be used on the Red Planet. Astronauts are able to train for what they will face when they get there.

HOW DO YOU NOT
GO CRAZY?
—ASTRONAUT CHRIS HADFIELD
ABOUT A POSSIBLE MISSION TO MARS, AS TOLD TO JOURNALIST ELMO KEEP

NASA has research sites around a meteor crater near Flagstaff, Arizona, where it conducts its Desert Research and Technology Studies (a.k.a. Desert RATS). The region is arid, windy, and rocky—perfect for testing equipment and training crews, but without quite as much of the inconvenience of more isolated locations. Among the ongoing projects is its planetary exploration geophysical system, or PEGS, which is mandated to develop seismic hardware suitable for use during Martian or lunar exploration. The system was also tested in Antarctica—another useful analog for Mars.

NASA is not the only entity attempting to simulate Mars. The Mars Desert Research Station in Utah is part of a projected global program of Mars research stations intended to conduct field explorations in geology and biology under many of the same constraints that would be encountered on Mars. Run

[**NASA's Desert RATS** program has a website where the public can learn about the technology and studies behind the scenes. Find fact sheets, mission blogs, photographs, and more at NASA.gov.]

by the Mars Society—a group of scientists and other interested parties—the Mars Desert Research Station makes its mission reports available on its website and uses public outreach to foster interest in Mars.

Devon Island, an inconveniently located isle in Canada's High Arctic, is home to another Mars Society site, the Flashline Mars Arctic Research Station. The island is cold, rocky, barren, and far from the creature comforts of civilization. It is a place where equipment can be tested to see how well it copes with rocks and wind, and where astronauts can practice working in space suits and driving prototype rovers. The Mars Society runs simulated missions there to determine everything from the best crew size, to how long it takes to stop a rover, put on a space suit, take rock samples, and get back in the rover.

WHAT A DIFFERENCE A SOL MAKES

MARTIAN DAY AND EARTH DAY, MARTIAN YEAR AND EARTH YEAR

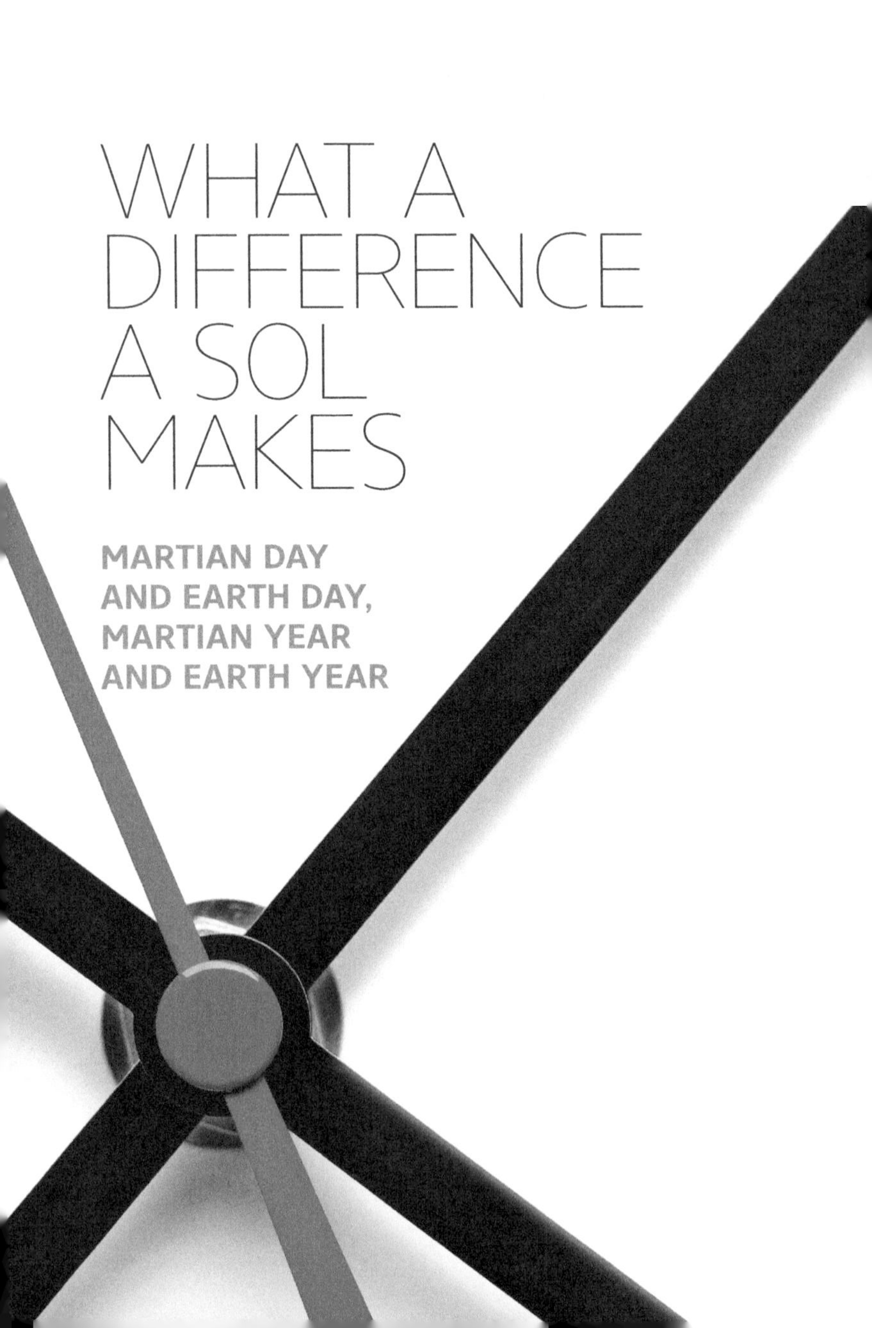

A day on Earth (more specifically, a solar day) is defined as the amount of time it takes for the planet to spin once around its own axis; in other words, it's how long it takes for the sun to return to the same position in the sky. The time from high noon with the sun directly overhead (at the equator, at least) to high noon again equals one day. We can think of a "Martian day" as the amount of time it takes for Mars to spin once on its axis, too, but to avoid confusion scientists use a different word for a Martian solar day. They call it a "sol."

An Earth day is 24 hours long, while a Martian sol is about 24 hours and 37 minutes. However, despite the similarities, the extra time of the sol can create problems for scientists on Earth. Keeping up with a sol schedule can induce a constant state of jetlag in researchers, and it's likely that explorers or colonists would face similar issues.

The Martian year is nearly twice as long as Earth's. While an Earth year is about 365.25 Earth days, a Martian year is 687 Earth days (we use Earth days to measure this in order to be able to directly compare the two; in Martian terms, Mars's year is 668.59 sols). That might not seem like a big deal, but remember that Mars has seasons very much like Earth's (though rather colder overall).

Mars has winter and summer, with transitional seasons in between, but each season lasts longer, with a more variable duration, than seasons on Earth. Imagine having to acclimatize to a winter that is eight months long, or more, instead of four.

HOME SWEET HABITAT

BUILDING A COLONY ON MARS

Mars has a lot going for it in terms of colonization. Its day duration and axial tilt are similar to Earth's, and its basic terrestrial nature makes it a good choice. But Mars has that pesky thin, poisonous atmosphere, high solar radiation, and low atmospheric pressure.

Pressurized living spaces would be a must, and any habitation would need to have facilities for manufacturing breathable oxygen (as well as oxygen for facilitating the burning of fuel) and a way to extract water from Mars's ground ice. Radiation shielding would be essential as well, because Mars does not have an ozone layer or enough of an atmosphere to provide any protection. Some researchers have suggested that any long-term Mars habitats should therefore be constructed underground, perhaps in existing caves or lava tubes, where the rocks and soil would act as a natural radiation shield.

A long-term colony or station on Mars would eventually need to produce food, fuel, water, breathable oxygen, and possibly even raw materials for further construction or repair. Recycling of all resources would be vital, so spacecraft, capsules, and other equipment would need to be designed to be repurposed once they arrive on the planet.

MODEL HOMES

NASA is working on mock-ups of a deep space habitat concept demonstrator modeled after the International Space Station. This deep space habitat is a modular design, with living

---

Left: Commander Astronaut Eugene A. Cernan makes a short checkout of the *Lunar Roving Vehicle* (*LRV*) during the first Apollo 17 extravehicular activity (EVA) at the Taurus-Littrow landing site, December 11, 1972.

FIRST WAVE

Most Mars colonization plans involve sending robotic landers and rovers to the Red Planet first, to set up basic facilities and begin production of fuel, water, and oxygen. When the first colonists and researchers arrive, they will already have the essentials for survival (and possibly even for a return to Earth) waiting for them.

components and various lab and other components added to rockets and Orion spacecraft to create a space-based station. Such a habitat is designed for use as space exploration moves out of low-Earth orbit toward the moon, asteroids, and eventually Mars.

The initial specifications for the habitat would allow a crew to live and work in space for up to a year and could include a multipurpose logistics module, or MPLM, an International Space Station–size lab-and-habitat element for living and working space, and a utility tunnel with an airlock and a space exploration vehicle (SEV) or other support craft for short exploration trips. When used in an actual space mission, the craft would also include a propulsion stage and an Orion spacecraft to transport crews to and from mission destinations.

One of the possible support craft for NASA's deep space habitat is the SEV. This small craft would be capable of carrying two crewmembers and would have spacesuits carried externally in a way that would allow astronauts to don the suit from inside the craft and then detach with minimal loss of gases. The SEV is being designed so that the same capsule can be mounted on a chassis with wheels or fitted out as space-based craft. The SEV design is based on the lunar electric rover, and the prototype is equipped with lithium-ion batteries.

DESERT RATS AND THE MARS SOCIETY

NASA has also been testing ground-based space habitats at its Desert RATS station in Arizona, using a modular capsule design that can have different types of research and work stations fitted to it, depending on the needs of the mission. This habitat consists of three primary elements: an entrance capsule designed to keep dust out of the main unit, a central main pod, and a hygiene module.

The main central element is circular with eight segments around a central lift, each of which can be swapped out. Some of the possible segments are a medical station, a geology lab, a maintenance station, and a communications center. The central lift gives access to an inflatable loft at the top of the capsule. NASA plans to equip the habitat with an intelligent software system, including sensors and automatic operations that would free astronauts for other tasks.

The Mars Society's desert and arctic research stations are also testing Mars habitat technology. Its design looks a lot like NASA's Desert RATS habitat from the outside, but inside it is less modular and more like a dormitory. Each station serves as the home and base of operations for four to six crew members on missions of several months' duration, simulating the types of conditions crews would face living and working in an off-Earth station.

FARMING THE RED PLANET

WHAT WILL THEY EAT?

Astronauts, like all humans, consume quite a large volume of food every year. But unlike most people, astronauts can't just pop out to the grocery store to restock the pantry. Everything a space mission's crewmembers will need to eat has to be brought with them, unless it can be grown.

Several space missions have included experiments to see how plants would respond to zero gravity, but the biggest concern isn't growing food for short-term spaceflights—it's for long-term research stations and colonies on the surfaces of other planets. Such missions would stay in one place long enough that greenhouses or hydroponic modules would be practical. But even greenhouses need sources of soil, water, and sunlight (or suitable artificial light), and these are things that would be impractical to bring along in large quantities.

Sunlight, at least, isn't likely to be too much of an issue. Although Mars is farther from the sun and thus receives less intense sunlight, it has nearly the same day length we do and receives a light spectrum similar to that we get on Earth. With the help of greenhouses or other contained units to boost warmth, plants native to Earth could probably manage on Mars levels of natural sunlight. Water, too, isn't too much of a problem.

Model of future Mars base with two cylindrical landing vehicles and greenhouse areas.

It would have to be melted and extracted from the ground, but Mars does have water. Even if it is salty, hard, or otherwise not perfectly usable as is, we already know filtration methods that don't require big advances in technology.

Soil might initially seem to be a bigger hurdle. Earth plants are accustomed to growing in Earth soils, after all. But soil analysis from NASA's *Phoenix* lander showed that at least some Martian soils are similar enough to Earth soils that growing edible plants in them is within the realm of possibility.

In fact, Earth lichens have been shown to be adaptable to Marslike conditions. In an experiment conducted by the German Aerospace Center, Antarctic lichens spent 34 days in the Mars Simulation Facility-Laboratory. Although they showed a preference for growing in cracks in the rocks and gaps deeper in the soil—areas where they were more sheltered from radiation than on the surface—the lichens otherwise carried on life as normal. Antarctic lichens are extremophiles, having already adapted to harsh environments, but if they can survive the Martian atmosphere unaided, perhaps less hardy plants could flourish in a protected Martian greenhouse. The real issue with Martian soil is not its chemical composition, but its lack of organic content. Earth soils contain as much—or even more—decomposing plant matter (as well as animal matter, bacteria and microbes, and other constituents) as they do rocks, dirt, and mineral matter. These organic components are vital to flourishing plants, which is why avid gardeners are willing to spend so much time and money on good fertilizer. Plants in sterile soil require feeding, not just water, and while it's possible to use chemical fertilizers, every farmer knows manure is worth its weight in gold. On Mars, manure might be worth considerably more than gold, because no one would have much use for pretty jewelry (gold is used in electronics and other space technology, but in very small amounts).

Taking all that into account, growing plants on Mars is probably quite possible. Which plants would do well under those conditions, on the other hand, is less certain. Plus, it is still to be determined whether the nutritional requirements of a crew of astronauts or a settlement of colonists could be met by Martian farming.

RED PLANET GREENHOUSES

While human settlers on Mars would need to live in pressurized habitats, it would be inefficient to pressurize enormous greenhouses to the same extent. The closer greenhouse pressures can be kept the same inside and outside, the easier they will be to build and maintain.

To that end, plant scientists at NASA are studying the effects of low-pressure environments on plant growth and have made some interesting discoveries. For instance, they found that even when they have more than enough water, plants react to very low-pressure environments the same way they react to drought; this is because the low pressure draws water out of their leaves more quickly.

PRIVATE GREENHOUSES

Though NASA and other government space agencies have been the biggest players in space exploration, especially in the development of spacecraft, other parties are becoming more significant.

The Mars Society has two Mars simulator projects making advances in testing technology and habitats for use on the Red Planet. As part of its Mars Desert Research Station in Utah, the Mars Society is also testing a "GreenHab"—a greenhouse module for growing food that also recycles the habitat module's water so nothing goes to waste.

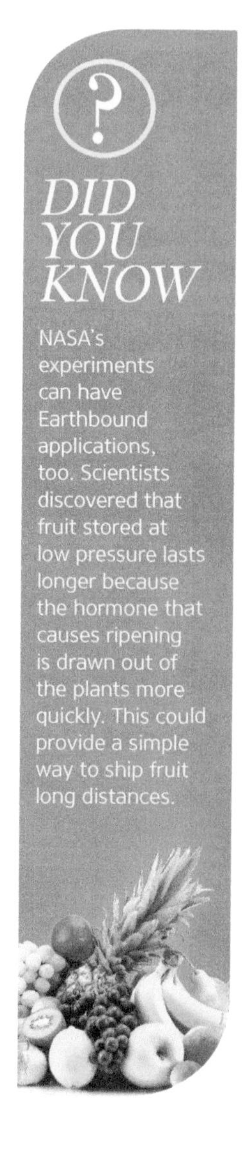

?
DID YOU KNOW
NASA's experiments can have Earthbound applications, too. Scientists discovered that fruit stored at low pressure lasts longer because the hormone that causes ripening is drawn out of the plants more quickly. This could provide a simple way to ship fruit long distances.

Billionaire Elon Musk and his SpaceX spaceflight company aim to establish a Mars colony, too. Musk's idea is to have space colonists buy a place in the Mars colony like a person might buy a house and property on Earth. Securing a single spot in Musk's proposed Mars colony would cost a cool $500,000. SpaceX's Mars settlement program aims to be largely self-sufficient, and plans include equipment to extract or produce oxygen, methane, and fertilizer from Mars's atmosphere and soil. Transparent domes pressurized with CO_2 from Mars's atmosphere would become greenhouses for growing Earth plants.

EXTRA-TERRESTRIAL ETHICS

CONSIDERING THE MOTIVES AND RESPONSIBILITIES OF SPACE EXPLORATION

Are there ethical considerations for space exploration in general and for the colonization of Mars in particular? After all, we have yet to encounter anything else living in space, so who would get hurt if we were to take a first-come, first-served approach and exploit whatever resources there were to be found?

Except for the presence or absence of life, the same might be said of Antarctica, but that continent is ruled by no entity on Earth but science—for the benefit of all nations.

Above: Kobus Vermeulen waters his garden in his self-made space suit. Vermeulen was shortlisted as a candidate to take part in the Mars One mission.

And if there is other life out there somewhere in the universe, shouldn't we share instead of claiming everything for ourselves?

LANGUAGE LESSON

In Latin, the language of science, the proper name of our star, the sun, is sol. *Hence why we talk about solar years and solar sails—things having to do with the sun are called "solar." Similarly, our moon is called* luna *in Latin, which is where we derived the adjective "lunar."*

NONPROFIT SPACE EXPLORATION

Most government space agencies can be considered nonprofits as far as space exploration is concerned. Certainly, they might have less-than-pure motives, as did the United States and USSR during the Cold War, when the Space Race was a handy tool of propaganda and a way to assert one country's superiority over the other. But in general, government space agencies fund space exploration for the sake of exploration itself. If they happen to gain technologies that can be used elsewhere or discover valuable new resources along the way, so much the better.

But there are also some nonprofit organizations in the field that exist to celebrate the science and make discoveries on behalf of all humankind. Former NASA scientist Robert Zubrin is founder and president of the Mars Society, which has members worldwide and funds projects by scientists from a variety of nations and disciplines. The society has grown and now welcomes those who aren't scientists to join and apply to participate in its projects, such as Mars Direct. Zubrin created the Mars Direct project to bring scientific minds to bear on the project of getting humans into space and settled on Mars. He argues that we should go to Mars (and beyond) not only because we can, but also because we must if we are to survive as a species.

Another nonprofit, Mars One, claims it will send people to Mars within the next decade or so to spread our reach out into the solar system; it has already accepted thousands of applications for a one-way trip to establish a colony. Despite the idealism of Mars One's founders, the organization has been criticized for inconsistencies in its reporting and documentation. Closer investigation revealed that Mars One has few or no ties with any of the providers of equipment that they claimed, and the number of applicants was far fewer than reported, among other concerns.

FOR-PROFIT SPACE EXPLORATION

There are undoubtedly useful resources in space, including some that are in short supply here on Earth. Why not mine asteroids for useful metals, for example? Or why not take tourists into low-Earth orbit at a ticket price of many thousand dollars per head? Companies have been created to take advantage of the new possibilities opened up by space travel, and the ethical implications of our actions must be considered.

This is not to suggest that all for-profit companies have ill intentions where space travel is concerned. These companies can do useful work and develop technologies that will eventually be available to everyone (due to limited time frames on patents). Spaceport America, for example, is not exploring space itself, but is providing a site for other companies to store and launch their crafts. And arguably, anyone who wishes to make a living in an industry fueled by space travel and exploration should be free to do so—within reason, and while adhering to ethical guidelines.

Both SpaceX and Virgin Galactic have been hard at work developing new spacecraft, and though naturally they expect to make a profit, they can also do good work for humankind at the same time. According to promotional material on their website, SpaceX aims to "revolutionize space technology with the ultimate goal of enabling people to live on other planets," while Virgin Galactic wants to "make space accessible to more people and for more purposes than ever before." At this early stage, it's impossible to know what the impacts and implications of for-profit space exploration might be—except that it's certainly excellent food for thought.

MARS TWO

MARS ONE'S TWO FOUNDERS

Mars One was founded by Bas Lansdorp and Arno Wielders

Arno Wielders earned a Master's degree in physics from the Free University of Amsterdam in 1997, then worked at Leiden Observatory. After some further graduate research, Wielders worked on NASA's ozone monitoring instrument program, then founded Space Horizon in 2005. In 2011 he cofounded Mars One and became its chief technical officer.

Bas Lansdorp (above) is a Dutch entrepreneur. He earned his Master's degree in mechanical engineering in 2003 from the University of Twente, then worked at Delft University of Technology before founding the wind energy company Ampyx Power in 2008. In 2011 he cofounded Mars One and became its CEO.

INSIDE THE MARTIAN MIND

WHAT DOES IT TAKE TO MAKE A ONE-WAY JOURNEY TO MARS?

In the early days of the space race, NASA's astronauts were selected from among experienced test pilots. The ideal candidate could remain calm in dangerous situations, and if they performed well for the TV cameras, that was even better. They were selected, as author Mary Roach points out in her book *Packing for Mars*, for their "balls and charisma."

Daring was important when first venturing into space; space travel was, and remains, a dangerous proposition. But after the moon landing, when the space program was focused more on conducting experiments in low-Earth orbit, and travel there and back was more routine, the necessary qualifications changed. Instead of daredevils, the space program needed scientists. Pilots

---

Left: Russian cosmonaut Alexander Samoukutyaev prepares to immerse in a swimming pool in a spacesuit as part of a training session at the Russian cosmonaut training facility in Star City outside Moscow.

DID YOU KNOW

One of the tests potential astronauts for Japan's JAXA space agency must undergo is to spend a week in an isolation facility with other finalists under strictly regimented conditions. The candidates are given a variety of tasks to complete, such as folding 1,000 paper origami cranes in a limited time period. The paper-crane folding tests speed, accuracy, and consistency when performing repetitive tasks.

were still needed to fly the craft, of course, but most of the crew were to be biologists, engineers, or chemists instead.

Now that the international community of space researchers is turning its attention to deeper space, including Mars and beyond, the desirable qualifications will change yet again, expanding to include individuals with varied skill sets. And the longer duration of planned missions means that the psychological characteristics of potential colonists will be more important than ever.

Some psychological issues facing humans in space are already known from previous missions. "Space euphoria," for example, can strike a spacewalking astronaut. This manifests as a feeling of overwhelming rightness and invulnerability, making him or her extremely reluctant to return to the spacecraft. It is similar to a condition among deep sea divers called "raptures of the deep," in which depths greater than 100 feet induce nitrogen narcosis; this results in effects similar to euphoric drunkenness.

Space euphoria has been much less of an issue, however, than extravehicular activity (EVA) height vertigo, which is also caused by the sight of Earth passing below a spacewalking astronaut. Instead of feeling invincible, as one might feel when experiencing space euphoria, EVA creates a feeling of falling at a high speed toward the planet below. Technically, this is actually what's happening—it's what being in orbit means: falling towards a planet and missing. EVA height vertigo and its resulting fear is a normal response to the situation, so it's not possible to screen out potential astronauts who might get it. Virtual reality training can help, but in the end, astronauts have to learn how to deal with it.

Perhaps more significant than how they deal with conditions outside the spacecraft is how candidates for long-term space exploration deal with the conditions inside the craft. Space vehicles are by necessity cramped, and even a habitat on the surface of Mars will have to be a small, enclosed space. There won't be a lot of room to get away from the other crewmembers, or even to stretch one's legs.

JAXA's isolation test is designed partly to see how potential astronauts fare when shut up with only a few others for constant company, but the duration of that test is only a week. Longer-term tests are much more useful, not only to test potential candidates, but also to look for general clues about how people will interact under the conditions Mars colonists and explorers would face.

To that end, Russia's Roscosmos and the Institute of Biomedical Problems in Moscow teamed up with China and the European Space Agency to conduct a series of tests during a project called Mars500. As part of this project, a facility at IBMP was constructed to simulate a complete space voyage, including launch and landing craft as well as habitat modules and a simulated Mars surface.

For the first test, a 14-day simulation in November 2007 of Mars living conditions tested the facility and procedures. This was followed by a 105-day study in 2009 of how four Russian and two European crew members would cope with the isolation. Finally, the full 520-day mission began on June 3, 2010, with three Russian, one Chinese, and two European crewmembers; it was conducted exactly as if it were a real mission, complete with simulated liftoff. The Mars500 project concluded on November 4, 2011, with a simulated landing.

During the course of the 520 days, the crew was subjected to a variety of tests to monitor their health and well-being. Initial conclusions suggested that four of the six crew members showed signs of problems that could have become problematic during a real mission, but nothing necessitated shutting down the experiment. The participants' sleep cycles were affected, partly because they were bored, as there wasn't a lot for them to do.

Like other tests and developments related to space exploration, Mars500 has produced results that can be of use to non-space travelers as well. Studies of astronauts' sleep cycles, for example, have applications to many earthbound situations.

TO TERRAFORM, OR NOT TO TERRAFORM?

THE POSSIBILITIES OF MAKING THE RED PLANET TURN GREEN

Terra is the name for Earth in Latin, so terraforming is literally "making Earth" out of another planet. There are many different possible processes for terraforming, all of them large-scale by necessity. Some are more labor-intensive than others, but all of them would take a long time. A very, very long time.

Above: The first phase of terraforming Mars would involve raising the temperature (from today's average of around -60 degrees Celcius) so that water remains liquid. The planet could then be oxygenated by plants. Such a project would take centuries to complete.

WHAT . . .

Most plans for terraforming Mars begin with methods for inducing the greenhouse effect. On Earth, we think of the greenhouse effect as a bad thing; it is at the root of our global climate change, after all. But the greenhouse effect is also what keeps the surface of Earth warm enough for our comfort. On Mars, the global temperature would need to rise if the planet were ever to support human life without the use of special habitats.

Mars is right on the edge of the Goldilocks zone of habitable space around our sun. It is so close to the edge, in fact, that during large parts of its orbit, it slips outside the zone. Warming the surface would be vital. The easiest way to do that would be by releasing the CO_2 stored in Mars's polar ice caps to create a blanket of greenhouses gases that could help retain the sun's heat. Mars's low gravity would prove problematic, because the gases would have the tendency to drift off into space instead of staying just above the surface.

Mars's polar caps are composed of both water ice and "dry" ice (frozen carbon dioxide). Mars does not have the atmospheric pressure needed to maintain either water or CO_2 in liquid form. When when it gets warm enough, the ice in the polar caps sublimates instead of melting—that is, both water and carbon dioxide go directly from solid to gaseous form.

Thickening the atmosphere of Mars by releasing greenhouse gases would also have the advantage of increasing atmospheric

[**Mars has seasons** and its own weather system—high white clouds, ground mists and frosts, and planet-wide dust storms that last for months. Terraforming would, of necessity, change or destroy much of this.]

pressure. Higher pressure would help Earth plants grow more normally (see the section on Farming the Red Planet for more on how low atmospheric pressure affects plants), and would allow water to exist on the surface in liquid form. Inducing the greenhouse effect on Mars would return the planet to conditions it likely had in the distant past—those of a warmer, wetter planet capable of supporting life.

Once the Red Planet is pressurized, so to speak, humans would not need spacesuits or pressurized habitats, though we would still need oxygen tanks. Eventually, perhaps, plant life that flourishes in CO_2 (because most plants "breathe" in carbon dioxide and breathe out oxygen) might fill the atmosphere with enough oxygen to make Martian air breathable to humans and other Earth animal life. And an oxygenated atmosphere would create an ozone layer, too, which would help protect the planet's inhabitants from the sun's harmful ultraviolet rays.

HOW . . .

The real issue isn't what to do to make Mars habitable, it's about how to do it.

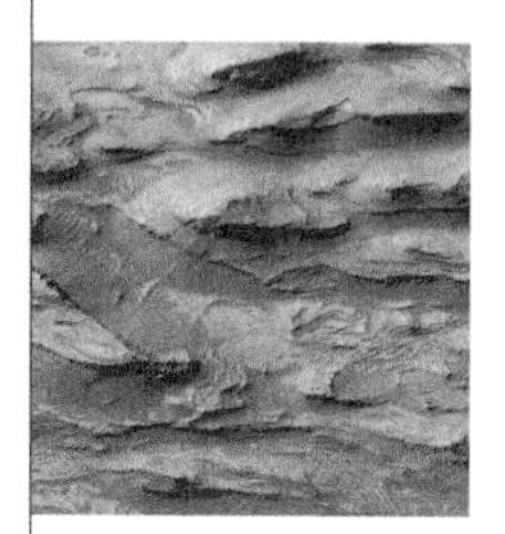

NOVEL WORLDS

Novels that show the terraforming of Mars include The Sands of Mars *by Arthur C. Clarke (1951),* The Martian Way *by Isaac Asimov (1952),* The Greening of Mars *by James Lovelock and Michael Allaby (1984),* Mining the Oort *by Frederik Pohl (1992), and Kim Stanley Robinson's* Mars *series, beginning with* Red Mars *(1992 to 1999).*

How do we get carbon dioxide from the polar caps and under the soil to sublimate and form an atmosphere? It would have to be done rapidly enough so the newly released CO_2 wouldn't just drift away into space before it could accumulate.

Heat is the key. But if we expect the greenhouse effect to do the warming, then we've got things backward. We have to apply heat to melt the ice to form the atmosphere to warm the planet. It seems circular. But there are a number of other possibilities for heating up the polar ice. One suggestion is to darken the polar ice caps, perhaps with carbon from a broken-up asteroid. The dark color would absorb more sunlight, in turn warming the polar ice so it melts. A similar idea would be to introduce dark-colored microbes or plant life to the poles. This life would have to be capable of surviving in extreme conditions. We know that Earth lichens could survive, but they tend to prefer cracks and crevices and spots away from direct radiation rather than hanging out right on the surface.

Another possibility is gigantic space mirrors to reflect more sunlight onto the poles. Or we could install factories on the surface of Mars and let them spew their greenhouse gases into the Martian sky. What was bad news for Earth might actually work in our favor on Mars. The factories could stay busy manufacturing useful items for Mars colonists or incinerating their waste.

BUT . . .

Perhaps a better question than, "Can we terraform Mars?" would be, "Should we terraform Mars?" Science fiction writer Ben Bova thinks the answer to the latter question is no. Though he supports missions to Mars, he believes we should keep the planet intact. Mars is Mars, and if we simply make it a new Earth, we would obliterate what it once was and might unwittingly destroy something valuable.

But what if Earth were destroyed and we needed a whole new home planet? Perhaps terraforming Mars would be the only way. It's more likely, though, that something that could destroy Earth (beyond our own stupidity, perhaps) would also destroy Mars. We could certainly make Mars an outpost on our way to a wider universe—where we might find earthlike planets already waiting for us—without changing the very nature of the Red Planet.

A PARTING THOUGHT

"Studying whether there's life on Mars or studying how the universe began, there's something magical about pushing back the frontiers of knowledge. That's something that is almost part of being human, and I'm certain that will continue."

—SALLY RIDE, AMERICA'S FIRST WOMAN IN SPACE

BIBLIOGRAPHY

Allison, Michael. "Telling Time on Mars." January 1988. NASA Goddard Institute for Space Studies. Accessed March 17, 2015. www.giss.nasa.gov/research/briefs/allison_02/

Anthony, Sebastian. "The First Mars One Colonists Will Suffocate, Starve, and be Incinerated, According to MIT." ExtremeTech. October 13, 2014. Accessed March 19, 2015. www.extremetech.com/extreme/191862-the-first-mars-one-colonists-will-suffocate-starve-and-be-incinerated-according-to-mit

ASI official website. Accessed March 12, 2015. www.asi.it/en

Baldwin, Emily. "Lichen survives harsh Mars environment." Skymania News, 26 April 2012. Accessed March 18, 2015. www.skymania.com/wp/2012/04/lichen-survives-harsh-martian-setting.html

BBC News. "Martian Soil 'Could Support Life'." June 27, 2008. Accessed March 18, 2015.

Beagle 2 Official Site. Accessed March 15, 2015. www.beagle2.com

Bova, Ben. "Afterword: The Once and Future Mars," in *The War of the Worlds: Mars' Invasion of Earth, Inciting Panic and Inspiring Terror from H. G. Wells to Orson Welles and Beyond*. Naperville, IL: Sourcebooks, 2005. Pages 255-261.

Bradbury, Ray. *The Martian Chronicles*. New York: Spectra, 1950.

Brandon, John. "7 Gadget Predictions Sci-Fi Authors Got Right." Popular Mechanics. Accessed March 12, 2015. www.popularmechanics.com/culture/g1232/7-gadget-predictions-sci-fi-authors-gor-right/

Brewer, E. Cobham. *Dictionary of Phrase and Fable*. 1898. Accessed March 15, 2015. www.bartleby.com/81/11799.html

Broad, William J. "Wreckage in the Desert Was Odd but Not Alien." *The New York Times* September 18, 1994. Accessed March 16, 2015. www.nytimes.com/1994/09/18/us/wreckage-in-the-desert-was-odd-but-not-alien.html

Brown, Eryn. "Is Exploring Mars Worth the Investment?" Los Angeles Times, July 30, 2012. Accessed March 19, 2015. articles.latimes.com/2012/jul/30/science/la-sci-mars-science-cost-20120730

Canadian Space Agency. "Canadarm." Accessed March 13, 2015. www.asc-csa.gc.ca/eng/canadarm/

Canadian Space Agency Official Site. Accessed March 12, 2015. www.asc-csa.gc.ca/eng/

Centre National d'Études Spatiales official website. Accessed March 12, 2015. www.cnes.fr/web/CNES-en/7114-home-cnes.php
Chandrasekhar, Indu. "Mars Curiosity Rover: Timeline of Discoveries." The Telegraph, December 16, 2013. Accessed March 15, 2015. www.telegraph.co.uk/news/science/space/10200893/Mars-Curiosity-Rover-timeline-of-discoveries.html
China National Space Administration. Accessed March 12, 2015. www.cnsa.gov.cn/n615709/cindex.html
Christou, Apostolos. "Astronomical Phenomena From Mars." Armagh Observatory. Accessed March 16, 2015. www.arm.ac.uk/~aac/mars/Information.html
Clute, John and Peter Nicholls, eds. *The Encyclopedia of Science Fiction*. New York: St. Martin's, 1995.
Coppinger, Rob. "Huge Mars Colony Eyed by SpaceX Founder Elon Musk." Space.com. November 23, 2012. Accessed March 18, 2015. www.space.com/18596-mars-colony-spacex-elon-musk.html
Cowling, Keith. "NASA has a Problem Calculating—and Admitting—What Space Missions Really Cost." Space Ref, June 3, 2005. Accessed March 19, 2015. www.spaceref.com/news/viewnews.html
Däniken, Erich. *Chariots of the Gods? Unsolved Mysteries of the Past*. New York: Souvenir Press Ltd, 1969.
Dayal, Geeta. "Music for Mars: 10 Songs to Celebrate Curiosity's Epic Landing." *Wired*, August 6, 2012. Accessed march 16, 2015. www.wired.com/2012/08/music-mars-rover-landing/
Diamond, Jared. *Guns, Germs, and Steel: The Fates of Human Societies*. New York: W.W. Norton, 1997.
Dismukes, Kim. "Timeline of Shuttle-Mir." NASA. Accessed March 12, 2014. spaceflight.nasa.gov/history/shuttle-mir/history/h-timeline.htm
DLR Official Site. Accessed March 12, 2015. www.dlr.de/dlr/en/desktopdefault.aspx/tabid-10002/
Do, Sydney, et al. "An Independent Assessment of the Technical Feasibility of the Mars One Mission Plan." 65th International Astronautical Congress. Accessed March 19, 2015. web.mit.edu/sydneydo/Public/Mars%20One%20Feasibility%20Analysis%20IAC14.pdf
ESA. "Robotic Exploration of Mars." Accessed March 11, 2015. exploration.esa.int/mars/
———. "ExoMars: ESA and Roscosmos set for Mars Missions." March 14, 2013. ESA Space Science. Accessed March 15, 2015.

www.esa.int/Our_Activities/Space_Science/ExoMars_ESA_and_Roscosmos_set_for_Mars_missions

———. "ESA's Participation in Mars500." Accessed March 19, 2015. www.esa.int/Our_Activities/Human_Spaceflight/Mars500

European Space Agency official website. Accessed March 12, 2015. www.esa.int/ESA

Ezell, Edward Clinton and Linda Neuman Ezell. *On Mars: Exploration of the Red Planet 1958-1978*. Accessed March 9, 2015. NASA History Office. Last updated August 9, 2004. history.nasa.gov/SP-4212/on-mars.html

FactMonster. "Timeline: Famous Firsts in Space Exploration." Accessed March 13, 2015. www.factmonster.com/science/astronomy/space-firsts-timeline.html

Freitas, Robert A. "Science and Science Fiction." *In Xenology: An Introduction to the Scientific Study of Extraterrestrial Life, Intelligence, and Civilization*. Sacramento, CA: Xenology Research Institute, 1979. Accessed March 13, 2015. http://www.xenology.info/Xeno/2.4.htm

"Galileo: the Telescope & the Laws of Dynamics." Astronomy 161 Web Syllabus, Department of Physics and Astronomy, University of Tennessee. Accessed March 17, 2015. csep10.phys.utk.edu/astr161/lect/history/galileo.html

Gaudin, Sharon. "Ice. Mineral-Rich Soil Could Support Human Outpost on Mars." June 28, 2008. IDG News Service. Accessed March 18, 2015.

Globus, Al. "Space Settlement Basics." April 29, 2013. NASA Astrobiology. Accessed March 18, 2015. settlement.arc.nasa.gov/Basics/wwwwh.html

Greene, Nick. "Space Firsts." Accessed March 12, 2015. space.about.com/cs/basics/a/spacefirsts.htm

Grotzinger, John P. "Habitability, Taphonomy, and the Search for Organic Carbon on Mars." *Science* 343, no. 6169 (January 2014): 386-387. Accessed March 16, 2014. www.sciencemag.org/content/343/6169/386

Hamilton, Peter F. "Fiction Prediction: The Legacy of Science Fiction Writers," Tor.com, October 9, 2014. Accessed March 12, 2015. www.tor.com/2014/10/fiction-prediction-the-legacy-of-science-fiction-writers

Hartmann, William K. *A Traveler's Guide to Mars: The Mysterious Landscapes of the Red Planet*. New York: Workman, 2003.

Harvard-Smithsonian Center for Astrophysics. "Eight New Planets Found in 'Goldilocks' Zone." Press Release 2015-04, January 6, 2016. Accessed March 10, 2015. www.cfa.harvard.edu/news/2015-04

History Learning Site. "The Cold War." Accessed March 10, 2015. www.historylearningsite.co.uk/coldwar.htm

HistoryShots. "Race to the Moon." Accessed March 12, 2015. www.historyshotsinfoart.com/space/timeline.cfm

Holmsten, Brian and Alex Lubertozzi. "Martians, Moon Men, and Other Close Encounters." *In The War of the Worlds: Mars' Invasion of Earth, Inciting Panic and Inspiring Terror from H. G. Wells to Orson Welles and Beyond*. Naperville, IL: Sourcebooks, 2005

Horton, Matthew. "Planet Pop: What Are the Best Songs About Mars?" NME. August 12, 2007. Accessed March 16, 2015. www.nme.com/blogs/nme-blogs/planet-pop-what-are-the-best-songs-about-mars

HubbleSite. "Why Do Astronomers Study Galaxies in Ultraviolet Light?" HubbleSite FAQ. Accessed March 17, 2015. hubblesite.org/reference_desk/faq/answer.php.id=42&cat=galaxies

IBN Live. "India Plans Second Mars Mission in 2018." October 24, 2014. Accessed March 15, 2015. ibnlive.in.com/news/india-plans-second-mars-mission-in-2018/509390-11.html

IMDb. "Ancient Aliens (2009-)." Accessed March 16, 2015. www.imdb.com/title/tt1643266/

———. "Unsolved Mysteries (1987-)." Accessed March 16, 2015. www.imdb.com/title/tt0094574/

IMDb. "Marvin the Martian (Character)". Accessed March 15, 2015. www.imdb.com/character/ch0030547/

Indian Space Research Organization. Accessed March 12, 2015. www.isro.gov.in

International Astronomical Union (IAU) Working Group for Planetary System Nomenclature (WGPSN). "Categories for Naming Features on Planets and Satellites." Gazetteer of Planetary Nomenclature. USGS website. Accessed March 15, 2015.

Japan Aerospace Exploration Agency official website. Accessed March 12, 2015. global.jaxa.jp

Jensen, K. Thor. "10 Sci-Fi Predictions That Came True," *PC Magazine*. Accessed March 12, 2015. www.pcmag.come/slideshow/story/330495/10-sci-fi-predictions-that-came-true/

Kaufman, Mark. "Ancient Mars Had an Ocean, Scientists Say." *New York Times*, March 5, 2015. Accessed March 16, 2015. www.nytimes.com/2015/03/06/science/mars-had-an-ocean-scientists-say-pointing-to-new-data.html

Keep, Elmo. "Mars One Finalist Explains Exactly How It's Ripping Off Supporters." Mattter, March 16, 2015. Accessed March 17, 2015. medium.com/matter/mars-one-insider-quits-dangerously-flawed-project-2dfef95217d3

Keep, Elmo. "All Dressed Up For Mars and Nowhere to Go." Matter, November 9, 2014. Accessed March 17, 2015. medium.com/matter/all-dressed-up-for-mars-and-nowhere-to-go-7e76df527ca0

Kerr, Richard E. "New Results Send Mars Rover on a Quest for Ancient Life." *Science* December 9, 2013. Accessed March 16, 2015. news.sciencemag.org/chemistry/2013/12/new-results-send-mars-rover-quest-ancient-life

Liptak, Andrew. "Weekend Playlist: Songs About Mars." SF Signal, December 2, 2011. Accessed March 16, 2015. www.sfsignal.com/archives/2011/12/weekend_playlist_songs_about_mars/

Neal, Matt. "Musicology: Songs About Mars." *The Standard*, August 12, 2015. Accessed March 16, 2015. www.standard.net.au/story/203213/musicology-songs-about-mars/

Malzberg, Barry N. "The Fifties." Originally published in *The Engines of the Night*. New York: Doubleday, 1982. Library of America. Accessed March 13, 2015. www.loa.org/sciencefiction/why_malzberg.jsp

Mann, Adam. "Future Mars Astronauts May be Sleepy, Bored and Crabby." *Wired* January 07, 2013. Accessed March 19, 2015. www.wired.com/2013/01/sleep-problems-mars-500/

Mars-500 official website. Accessed March 19, 2015. mars500.imbp.ru/en/index_e.html

Mars Desert Research Station website. mdrs.marssociety.org

Mars One official website. Accessed March 18, 2015. www.mars-one.com

Mars Society Website. Accessed March 16, 2015. www.marssociety.org

Mars Society. Mars Desert Research Station website. Accessed March 16, 2015. mdrs.marssociety.org

Mars Society. "GreenHab." Mars Desert Research Station. Accessed March 18, 2015. mdrs.marssociety.org/greenhab

Miles, Kathy. "The Martian Sky: Stargazing from the Red Planet." StarrySkies.com. Accessed march 16, 2015. starryskies.com/The_sky/events/mars/opposition08.html

NASA official website. www.nasa.gov

———. "Mars Facts." NASA Mars Exploration. Accessed March 10, 2015. mars.jpl.nasa.gov/allaboutmars/facts

———. "Mars Exploration." Accessed March 9, 2015. mars.jpl.nasa.gov

———. "The Space Exploration Initiative." NASA History Program Office. Accessed March 10, 2015. history.nasa.gov/sei.htm

———. "Venus: Overview." Solar System Exploration: Planets. Accessed March 10, 2015. solarsystem.nasa.gov/planets/profile.cfm?Object=Venus

NASA. "International Space Station." Accessed March 12, 2015. www.nasa.gov/mission_pages/station/main/index.html

———. "Space Shuttle: Shuttle Basics." Accessed March 12, 2015. Return to Flight. www.nasa.gov/returntoflight/system/system_STS.html

———. "Mars Exploration Rovers." NASA Jet Propulsion Laboratory. Accessed March 11, 2015. mars.nasa.gov/mer/home/

———. Mars Reconnaissance Orbiter Website. Accessed March 15, 2015. www.nasa.gov/mission_pages/MRO/main/index.html

———. "Missions to Mars." Accessed March 12, 2015. www.nasa.gov/mission_pages/mars/missions/

———. "Phobos-Grunt." Accessed March 13, 2015. nssdc.gsfc.nasa.gov/nmc/spacecraftDisplay.do?id=2011-065A

NASA. "2001 Mars Odyssey." NASA Jet Propulsion Laboratory. Accessed March 15, 2015. mars.jpl.nasa.gov/odyssey/index.cfm

———. "Mars Pathfinder Frequently Asked Questions: Sojourner Rover." Last update April 10, 1997. Accessed March 15, 2015.

———. "NASA Announces Mars 2020 Rover Payload to Explore the Red Planet as Never Before." NASA Press release 14-208, July 31, 2014. Accessed March 15, 2015. www.nasa.gov/press/2014/july/nasa-announces-mars-2020-rover-payload-to-explore-the-red-planet-as-never-before/

———. "Desert Research and Technology Studies (Desert RATS)." February 6, 2013. Accessed March 16, 2015. www.nasa.gov/exploration/human-research/analogs/research_info_analog-drats.html#.VQcDfykf8wI

———. "Constructing Next-Generation Space Habitat Demonstrators." July 16, 2012. Accessed March 18, 2015. www.nasa.gov/exploration/technology/deep_space_habitat/constructing-demonstrators.html

———. "Habitat Demonstration Unit—Deep Space Habitat." Lyndon B. Johnson Space Center. 2011. Accessed March 18, 2015. www.nasa.gov/pdf/468441main_HDU_FactSheet_508.pdf

———. “Habitation Systems Projects—Deep Space Habitat Facts.” Accessed March 18, 2015. www.nasa.gov/exploration/technology/deep_space_habitat/facts.html

———. “NASA’s Space Exploration Vehicle (SEV).” Beyond Earth. Accessed March 18, 2015. www.nasa.gov/exploration/technology/space_exploration_vehicle/index.html

———. “Greenhouses for Mars.” NASA Science: Science News. February 25, 2004. Accessed March 18, 2015.

———. “NASA Spacecraft Analyzing Martian Soil Data.” NASA Jet Propulsion Laboratory. August 4, 2008. Accessed March 18, 2015.

———. “Terraforming Mars.” November 25, 2006. Accessed March 19, 2015, via the Internet Archive. web.archive.org/web/20070915152013/http://aerospacescholars.jsc.nasa.gov/HAS/cirr/em/10/10.cfm

———. “Missions to Mars.” March 11, 2013. Accessed March 19, 2015. www.nasa.gov/mission_pages/mars/missions/

National Geographic. “Making Mars the New Earth.” *National Geographic Magazine*, January 15, 2010. Accessed March 19, 2015. ngm.nationalgeographic.com/big-idea/07/mars

O’Connor, J.J. and E.F. Robertson. “John Wilkins.” February 2002. School of Mathematics and Statistics, University of St Andrews, Scotland. Accessed March 10, 2015. www-history.mcs.st-andrews.ac.uk/Biographies/Wilkins.html

Oxford Dictionaries. “Mother Earth.” Accessed March 10, 2015. www.oxforddictionaries.com/definition/english/Mother-Earth

Palaeolexicon. “The Linear B Word Ma-Ka.” Accessed March 10, 2015. www.palaeolexicon.com/ShowWord.aspx?id=17016

Paliwoda, David and Jesse Williams. “How Far Is It to Mars?” Accessed March 10, 2015. www.distancetomars.com

Ride, Sally. BrainyQuote.com, Xplore Inc, 2015. http://www.brainyquote.com/quotes/quotes/s/sallyride589687.html. Accessed May 13, 2015.

Roach, Mary. *Packing for Mars: The Curious Science of Life in the Void*. New York: Norton, 2010.

Romano, Andrea. “15 Sci-Fi Books That Predicted the Future,” Mashable, July 23, 2014. Accessed March 12, 2015. mashable.com/2014/07/23/sci-fi-books-the-future/

Roberts, Adam. *The History of Science Fiction*. New York: Macmillan, 2005.

Robinson, Kim Stanley. *Red Mars*. New York: Spectra, 1993.

Russell, Randy. "A History of Satellites and Robotic Space Missions." Last updated May 20, 2008. Accessed March 20, 2015. www.windows2universe.org/space_missions/unmanned_table.html

Russian Federal Space Agency (Roscosmos) official website. Accessed March 12, 2015. en.federalspace.ru

RussianSpaceWeb.com. "Chronology: Moon Race." Accessed March 12, 2015. www.russianspaceweb.com/chronology_moon_race.html

Sanders, Robert. "Hundreds of Auroras Detected on Mars." December 12, 2005. Accessed March 16, 2015. phys.org/news8987.html

Saltarin, Alexander. "NASA Chief Defends Mars Mission, Colonization of Red Planet Necessary for Human Survival." *Tech Times*, April 24, 2014. Accessed March 18, 2015. www.techtimes.com/articles/5978/20140424/nasa-chief-defends-mars-mission-colonization-of-red-planet-necessary-for-human-survival.htm

Silverberg, Robert. "Science Fiction in the Fifties: The Real Golden Age." Originally published in 2010. Library of America. Accessed March 11, 2015. www.loa.org/sciencefiction/why_silverberg.jsp

Smithsonian. "Space Race." Smithsonian National Air and Space Museum. Accessed March 10, 2015. airandspace.si.edu/exhibitions/space-race/

Space.com. "Destination Mars: A Timeline of Red Planet Landings." July 9, 2012. Accessed March 15, 2015. www.space.com/16496-mars-landing-missions-timeline.html

——— "India's First Mars Mission in Pictures." Accessed March 13, 2015. www.space.com/23203-india-mars-orbiter-mission-photos.html

Space Pioneer Learning Adventures. "Sojourner." Robotic Rovers Home Page. Accessed March 15, 2014. spacepioneers.msu.edu/robot_rovers/sojourner.html

Spaceport America official website. Accessed March 18, 2015. spaceportamerica.com

SpaceX official website. Accessed March 18, 2015. www.spacex.com

Stableford, Brian. "The Third Generation of Genre Science Fiction." *Science Fiction Studies* 23 (3): 321–330 (November 1996).

StarPort Café. "Perception and the Overview Effect." 2011. Accessed March 19, 2015. www.starportcafe.com/space-background/perception-and-the-overview-effect

Stromberg, Joseph. "India's Mission to Mars Cost Less than the Movie Gravity." Vox, September 24, 2014. Accessed March 19, 2015. www.vox.com/2014/9/24/6838079/india-mars-mangalyaan

The Times of India. "China Unveils its Mars Rover After India's Successful 'Mangalyaan'." November 10, 2014. Accessed March 15, 2015. m.timesofindia.com/home/science/China-unveils-its-Mars-rover-after-Indias-successful-Mangalyaan/articleshow/45099803.cms

The Planetary Society. "Mars Exploration Rover Graphical Timeline." The Bruce Murray Space Image Library. Accessed March 15, 2015. www.planetary.org/multimedia/space-images/charts/mars-exploration-rover-grahical-timeline.html

The Planetary Society. "Space Missions." Accessed March 20, 2015. www.planetary.org/explore/space-topics/space-missions/

The Planetary Society. "Missions to Mars." Accessed March 19, 2015. www.planetary.org/explore/space-topics/space-missions/missions-to-mars.html

TheSpaceRace.com. "Timeline of Space Exploration." Accessed March 11, 2015. www.thespacerace.com/timeline/

Thornhill, Ted and Ellie Fagharifard. "India's Spacecraft Orbits Mars Successfully." Mail Online, September 24, 2014. Accessed March 19, 2015.

Virgin Galactic official website. Accessed March 18, 2015. www.virgingalactic.com

Wall, Mike. "Oxygen-Generating Mars Rover to Bring Colonization Closer." August 01, 2014. Space.com. Accessed March 18, 2015. www.space.com/26705-nasa-2020-rover-mars-colony-tech.html

Whitehouse, David. "Long History of Water and Mars." BBC News. January 24, 2004. Accessed March 16, 2015. news.bbc.co.uk/2/hi/science/nature/3426539.stm

William, David R. "Mars Fact Sheet." Accessed March 9, 2015. NASA Lunar and Planetary Science: Mars. Last updated February 27, 2015. nssdc.gsfc.nasa.gov/planetary/factsheet/marsfact.html

York, Paul. "The Ethics of Terraforming." *Philosophy Now,* 2002. Accessed March 19, 2015.

Zubrin, Robert M. and Christopher P. McKay. "Technological Requirements for Terraforming Mars." The Terraforming Information Pages. Accessed March 19, 2015. www.users.globalnet.co.uk/~mfogg/zubrin.htm

INDEX

N

O

P

R

S

T

U

V

CONTINUE THE
CONVERSATION

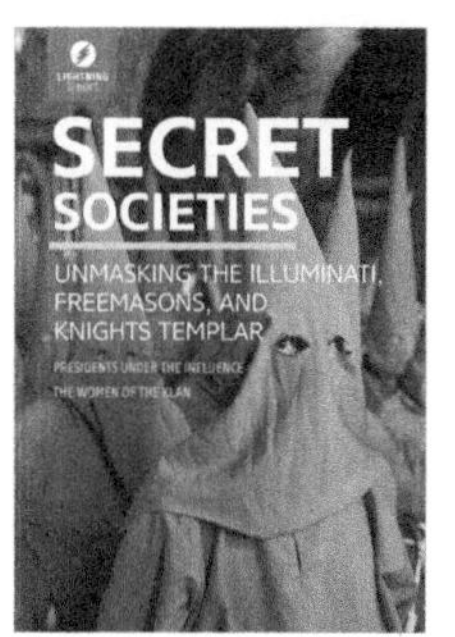

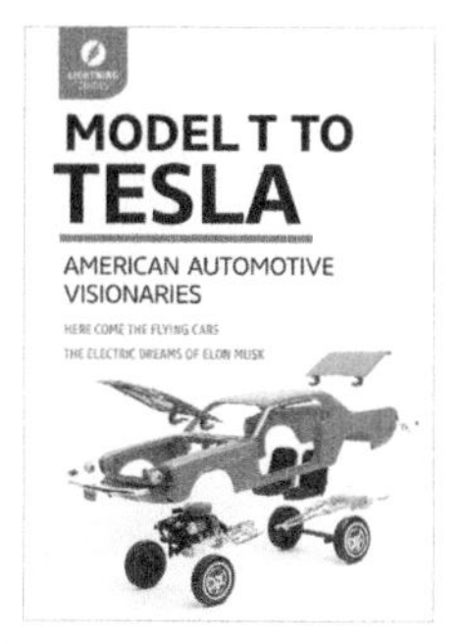

CPSIA information can be obtained at www.ICGtesting.com
Printed in the USA
BVOW11s0554290515

402313BV00003B/3/P